Seaplane, Skiplane, and Float/Ski Equipped Helicopter Operations Handbook

FAA-H-8083-23

U.S. DEPARTMENT OF TRANSPORTATION
FEDERAL AVIATION ADMINISTRATION
Flight Standards Service

Skyhorse Publishing

Skyhorse Publishing books may be purchased in bulk at special discounts for sales promotion, corporate gifts, fund-raising, or educational purposes. Special editions can also be created to specifications. For details, contact the Special Sales Department, Skyhorse Publishing, 307 West 36th Street, 11th Floor, New York, NY 10018 or info@skyhorsepublishing.com.

www.skyhorsepublishing.com

10 9 8 7 6 5 4 3 2 1

Library of Congress Cataloging-in-Publication Data available on file.

ISBN-13: 978-1-61608-202-4

Printed in China

PREFACE

This operational handbook introduces the basic skills necessary for piloting seaplanes, skiplanes, and helicopters equipped with floats or skis. It is developed by the Flight Standards Service, Airman Testing Standards Branch, in cooperation with various aviation educators and industry.

This handbook is primarily intended to assist pilots who already hold private or commercial certificates and who are learning to fly seaplanes, skiplanes, or helicopters equipped for water or ski operations. It is also beneficial to rated seaplane pilots who wish to improve their proficiency, pilots preparing for flights using ski equipped aircraft, and flight instructors engaged in the instruction of both student and transitioning pilots. It introduces the future seaplane or skiplane pilot to the realm of water operations and cold weather operations, and provides information on the performance of procedures required for the addition of a sea class rating in airplanes. Information on general piloting skills, aeronautical knowledge, or flying techniques not directly related to water or cold weather operations are beyond the scope of this book, but are available in other Federal Aviation Administration (FAA) publications.

This handbook conforms to pilot training and certification concepts established by the FAA. There are different ways of teaching, as well as performing specific operating procedures, and many variations in the explanations of operating from water, snow, and ice. This handbook is not comprehensive, but provides a basic knowledge that can serve as a foundation on which to build further knowledge. The discussion and explanations reflect commonly used practices and principles. Occasionally the word "must" or similar language is used where the desired action is deemed critical. The use of such language is not intended to add to, interpret, or relieve a duty imposed by Title 14 of the Code of Federal Regulations (14 CFR).

It is essential for persons using this handbook to also become familiar with and apply the pertinent parts of 14 CFR and the *Aeronautical Information Manual (AIM)*. The AIM is available online at **http://www.faa.gov/atpubs**. Performance standards for demonstrating competence required for the seaplane rating are prescribed in the appropriate practical test standard.

The current Flight Standards Service airman training and testing material and subject matter knowledge codes for all airman certificates and ratings can be obtained from the Flight Standards Service web site at **http://av-info.faa.gov**.

The FAA greatly appreciates the valuable assistance provided by many individuals and organizations throughout the aviation community whose expertise contributed to the preparation of this handbook.

This handbook supercedes Chapters 16 and 17 of FAA-H-8083-3, *Airplane Flying Handbook*, dated 1999. This handbook is available for download from the Flight Standards Service Web site at **http://av-info.faa.gov.** This Web site also provides information about availability of printed copies.

This handbook is published by the U.S. Department of Transportation, Federal Aviation Administration, Airman Testing Standards Branch, AFS-630, P.O. Box 25082, Oklahoma City, OK 73125. Comments regarding this handbook should be sent in e-mail form to **AFS630comments@faa.gov**.

AC 00-2, *Advisory Circular Checklist*, transmits the current status of FAA advisory circulars and other flight information and publications. This checklist is available via the Internet at **http://www.faa.gov/aba/html_policies/ac00_2.html.**

CONTENTS

Chapter 1

Rules, Regulations, and Aids for Navigation

PRIVILEGES AND LIMITATIONS

In general, the privileges and limitations of a seaplane rating are similar to those of the equivalent land rating. The same standards and requirements apply as for comparable landplane certificates.

While it is possible for a student to use a seaplane to obtain all the flight training necessary to earn a pilot certificate, and many pilots have done so, this publication is intended primarily for pilots who already hold airman certificates and would like to add seaplane capabilities. Therefore, this chapter does not address pilot certificate requirements, regulations, or procedures that would also apply to landplane operations. Information on regulations not directly related to water operations is available in other Federal Aviation Administration (FAA) publications.

For certification purposes, the term "seaplane" refers to a class of aircraft. A pilot requires additional training when transitioning to a seaplane. Ground and flight training must be received and logged, and a pilot must pass a class rating practical test prior to initial operations as pilot in command. This training requires the use of an authorized flight instructor to conduct such training and attest to the competency of a pilot prior to taking the practical test. Because the seaplane rating is part of an existing pilot certificate, the practical test is not as extensive as for a new pilot certificate, and covers only the procedures unique to seaplane operations. No separate written test is required for pilots who are adding seaplane to an existing pilot certificate.

Adding a seaplane rating does not modify the overall limitations and privileges of the pilot certificate. For example, private pilots with a seaplane rating are not authorized to engage in seaplane operations that would require a commercial certificate. Likewise, a pilot with a single-engine seaplane class rating may not fly multi-engine seaplanes without further training. However, no regulatory distinction is made between flying boats and seaplanes equipped with floats. [Figure 1-1]

SEAPLANE REGULATIONS

Because of the nature of seaplane operations, certain regulations apply. Most of them are set forth in Title 14

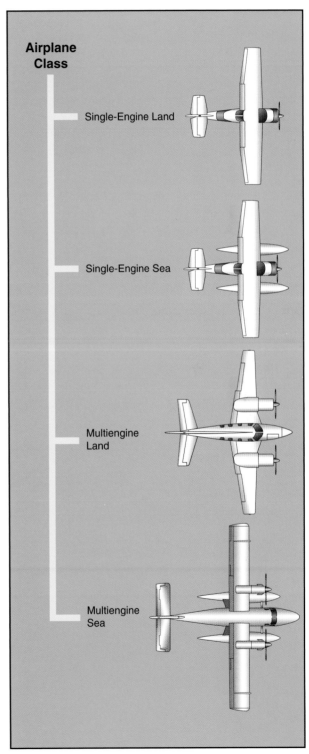

Airplane Class

Single-Engine Land

Single-Engine Sea

Multiengine Land

Multiengine Sea

Figure 1-1. Seaplane is a class.

of the Code of Federal Regulations (14 CFR) parts 1, 61, and 91.

Just as land-based pilots must understand airport operations, the pilot certification requirements in 14 CFR part 61 require seaplane pilots to know and use the rules for seaplane base operations.

Specific regulations recognize the unique characteristics of water operations. For example, 14 CFR part 61, section 61.31 takes into account that seaplanes seldom have retractable landing gear as such, so an endorsement to act as pilot in command of a complex seaplane requires training in a seaplane with flaps and a controllable pitch propeller. Likewise, in 14 CFR part 91, section 91.107, there is an exception to the rule that everyone must have a seat and wear a seatbelt during movement on the surface. The person pushing off or mooring a seaplane at a dock is authorized to move around while the seaplane is in motion on the surface.

14 CFR PART 91, SECTION 91.115 RIGHT-OF-WAY RULES: WATER OPERATIONS

The right-of-way rules for operation on water are similar, but not identical, to the rules governing right-of-way between aircraft in flight.

(a) General. Each person operating an aircraft on the water shall, insofar as possible, keep clear of all vessels and avoid impeding their navigation, and shall give way to any vessel or other aircraft that is given the right-of-way by any rule of this section.

(b) Crossing. When aircraft, or an aircraft and a vessel, are on crossing courses, the aircraft or vessel to the other's right has the right-of-way.

(c) Approaching head-on. When aircraft, or an aircraft and a vessel, are approaching head-on, or nearly so, each shall alter its course to the right to keep well clear.

(d) Overtaking. Each aircraft or vessel that is being overtaken has the right-of-way, and the one overtaking shall alter course to keep well clear.

(e) Special circumstances. When aircraft, or an aircraft and a vessel, approach so as to involve risk of collision, each aircraft or vessel shall proceed with careful regard to existing circumstances, including the limitations of the respective craft.

RULES OF THE SEA

According to United States Coast Guard (USCG) regulations, the definition of a vessel includes virtually anything capable of being used for transportation on water, including seaplanes on the water. Therefore, any time a seaplane is operating on the

water, whether under power or not, it is required to comply with USCG navigation rules applicable to vessels. Simply adhering to 14 CFR part 91, section 91.115 should ensure compliance with the USCG rules. Pilots are encouraged to obtain the USCG Navigation Rules, International-Inland, M16672.2D, available from the U.S. Government Printing Office. These rules apply to all public or private vessels navigating upon the high seas and certain inland waters.

INLAND AND INTERNATIONAL WATERS

Inland waters are divided visually from international waters by buoys in areas with frequent ocean traffic. Inland waters are inshore of a line approximately parallel with the general trend of the shore, drawn through the outermost buoy. The waters outside of the line are international waters or the high seas.

Seaplanes operating inshore of the boundary line dividing the high seas from the inland waters must follow the established statutory Inland Rules (Pilot Rules). Seaplanes navigating outside the boundary line dividing the high seas from inland waters must follow the International Rules of the Sea. All seaplanes must carry a current copy of the rules when operating in international waters.

UNITED STATES AIDS FOR MARINE NAVIGATION

For safe operations, a pilot must be familiar with seaplane bases, maritime rules, and aids to marine navigation.

SEAPLANE LANDING AREAS

The familiar rotating beacon is used to identify lighted seaplane landing areas at night and during periods of reduced visibility; however, the colors alternate white and yellow for water landing areas. A double white flash alternating with yellow identifies a military seaplane base.

On aeronautical charts, seaplane landing areas are depicted with symbols similar to land airports, with the addition of an anchor in the center. As with their land counterparts, tick marks around the outside of the symbol denote a seaplane base with fuel and services available, and a double ring identifies military facilities. [Figure 1-2]

BUOYS AND DAYBEACONS

Buoys are floating markers held in place with cables or chains to the bottom. Daybeacons are used for similar purposes in shallower waters, and usually consist of a marker placed on top of a piling or pole driven into the bottom. Locations of buoys within U.S. waters are

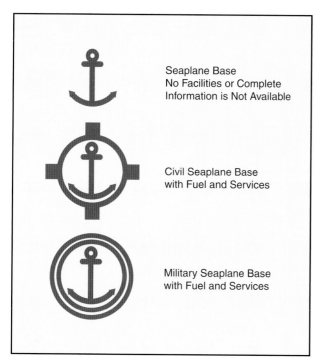

Seaplane Base
No Facilities or Complete
Information is Not Available

Civil Seaplane Base
with Fuel and Services

Military Seaplane Base
with Fuel and Services

Figure 1-2. Seaplane landing areas have distinctive symbols to distinguish them from land airports.

shown on nautical charts prepared by the Office of Coast Survey (OCS), an office within the National Oceanic and Atmospheric Administration (NOAA). Light lists prepared by the Coast Guard describe lightships, lighthouses, buoys, and daybeacons maintained on all navigable waters of the United States.

The buoyage system used in the United States employs a simple arrangement of colors, shapes, numbers, and lights. Whenever operating near buoys, keep in mind that the length of chain holding the buoy in place is likely to be several times the depth of the water, so the buoy may be some distance from its charted location, as well as from any danger or obstruction it is intended to mark. Do not come any closer to a buoy than necessary.

Buoys with a cylindrical shape are called can buoys, while those with a conical shape are known as nun buoys. The shape often has significance in interpreting the meaning of the buoy. [Figure 1-3]

Since a buoy's primary purpose is to guide ships through preferred channels to and from the open sea, the colors, shapes, lights, and placement become meaningful in that context. Approaching from seaward, the left (port) side of the channel is marked with black or green can buoys. These buoys use odd numbers whose values increase as the vessel moves toward the coast. They also mark obstructions that should be kept to the vessel's left when proceeding from seaward.

The right side of the channel, or obstructions that should be kept to the vessel's right when headed toward shore, are marked with red nun buoys. These

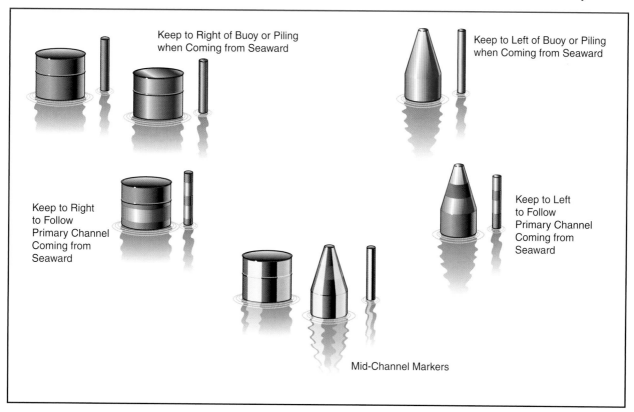

Keep to Right of Buoy or Piling
when Coming from Seaward

Keep to Left of Buoy or Piling
when Coming from Seaward

Keep to Right
to Follow
Primary Channel
Coming from
Seaward

Keep to Left
to Follow
Primary Channel
Coming from
Seaward

Mid-Channel Markers

Figure 1-3. Buoys typically used along waterways.

buoys use even numbers whose values increase from seaward. The mnemonic "red, right, returning" helps mariners and seaplane pilots remember to keep the red buoys to their right when proceeding toward the shore ("returning" to their home port).

Black and white vertically striped buoys mark the center of the channel or fairway (the nautical term for the navigable part of a river, bay, or harbor), and may use letters starting at A from seaward.

Naturally, not all waterways lead straight from ocean to port, so there are also buoys to mark the junctions of waterways. Buoys with red and black horizontal bands mark junctions or places where the waterway forks. They also mark wrecks and obstructions that can be passed on either side. The color of the top band (red or black) and the shape of the buoy (nun or can) indicate the side on which the buoy should be passed by a vessel proceeding inbound along the primary channel. If the topmost band is black, the buoy should be kept to the left of an inbound vessel. If the topmost band is red, keep the buoy to the right when inbound. Buoys with the black top band will usually be cans, while those with the red top band will usually be nuns.

For waterways that run more or less parallel to the coast, there is no obvious inbound or outbound to give direction to the waterway, so by convention the inbound direction of such waterways is assumed to be "clockwise" around the contiguous states. This means that for waterways running parallel to the east coast, southbound is considered the inbound direction; for waterways along the Gulf coast, inbound means westbound; and for waterways along the west coast, northbound is inbound.

Daybeacons and daymarks serve similar purposes as buoys and use similar symbology. In the United States, green is replacing black as the preferred color for port-side daymarks. [Figure 1-4]

These are just the most basic features of the most common buoyage system in the United States. There are other buoyage systems in use, both in the United States and in other countries. Sometimes the markings are exactly the opposite of those just described. Good pilots will obtain a thorough understanding of the maritime aids to navigation used in the areas where they intend to fly.

NIGHTTIME BUOY IDENTIFICATION
Usually only the more important buoys are lighted. Some unlighted buoys may have red, white, or green reflectors having the same significance as lights of the same colors. Black or green buoys have green or white lights; red buoys have red or white lights. Likewise, buoys with a red band at the top carry red lights, while those with a black band topmost carry green lights. White lights are used without any color significance. Lights on red or black buoys are always flashing or occulting. (When the light period is shorter than the dark period, the light is flashing. When the light is interrupted by short dark periods, the light is occulting.) A light flashing a Morse Code letter "A" (dot-dash) indicates a mid-channel buoy.

There is much more to the system of maritime navigation aids than can be presented here. Nautical books and online resources can be a great help in extending knowledge and understanding of these important aids.

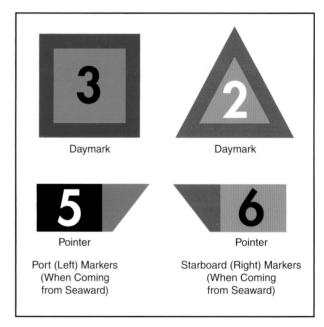

Figure 1-4. Typical daymarks.

Chapter 2

Principles of Seaplanes

SEAPLANE CHARACTERISTICS

There are two main types of seaplane: flying boats (often called hull seaplanes) and floatplanes. The bottom of a **flying boat**'s fuselage is its main landing gear. This is usually supplemented with smaller floats near the wingtips, called **wing or tip floats**. Some flying boats have **sponsons**, which are short, winglike projections from the sides of the hull near the waterline. Their purpose is to stabilize the hull from rolling motion when the flying boat is on the water, and they may also provide some aerodynamic lift in flight. Tip floats are sometimes known as sponsons. The hull of a flying boat holds the crew, passengers, and cargo; it has many features in common with the hull of a ship or boat. On the other hand, **floatplanes** typically are conventional landplanes that have been fitted with separate floats (sometimes called pontoons) in place of their wheels. The fuselage of a floatplane is supported well above the water's surface.

Some flying boats and floatplanes are equipped with retractable wheels for landing on dry land. These aircraft are called **amphibians**. On amphibious flying boats, the main wheels generally retract into the sides of the hull above the waterline. The main wheels for amphibious floats retract upward into the floats themselves, just behind the step. Additional training is suggested for anyone transitioning from straight floats to amphibious aircraft. [Figure 2-1]

There are considerable differences between handling a floatplane and a flying boat on the water, but similar principles govern the procedures and techniques for both. This book primarily deals with floatplane

Figure 2-1. Flying boats, floatplanes, and amphibians.

operations, but with few exceptions, the explanations given here also apply to flying boats.

A number of amphibious hull seaplanes have their engines mounted above the fuselage. These seaplanes have unique handling characteristics both on the water and in the air. Because the thrust line is well above the center of drag, these airplanes tend to nose down when power is applied and nose up as power is reduced. This response is the opposite of what pilots have come to expect in most other airplanes, and can lead to unexpected pitch changes and dangerous situations if the pilot is not thoroughly familiar with these characteristics. Pilots transitioning to a seaplane with this configuration should have additional training.

Many of the terms that describe seaplane hulls and floats come directly from the nomenclature of boats and ships. Some of these terms may already be familiar, but they have specific meanings when applied to seaplanes. Figures 2-2 and 2-3 describe basic terms, and the glossary at the end of this book defines additional terms.

Other nautical terms are commonly used when operating seaplanes, such as port and starboard for left and right, windward and leeward for the upwind and downwind sides of objects, and bow and stern for the front and rear ends of objects.

Research and experience have improved float and hull designs over the years. Construction and materials have changed, always favoring strength and light weight. Floats and hulls are carefully designed to optimize hydrodynamic and aerodynamic performance.

Floats usually have bottoms, sides, and tops. A strong **keel** runs the length of the float along the center of the bottom. Besides supporting the seaplane on land, the keel serves the same purpose as the keel of a boat when the seaplane is in the water. It guides the float in a straight line through the water and resists sideways motion. A short, strong extension of the keel directly behind the step is called the **skeg**. The **chine** is the seam where the sides of the float are joined to the bottom. The chine helps guide water out and away from the float, reducing spray and helping with **hydrodynamic** lift. Hydrodynamic forces are those that result from motion in fluids.

On the front portion of the float, midway between the keel and chine, are the two **sister keelsons**. These longitudinal members add strength to the structure and function as additional keels. The top of the float forms a **deck** that provides access for entering and leaving the cabin. Bilge pump openings, hand hole covers, and cleats for mooring the seaplane are typically located along the deck. The front of each float has a rubber bumper to cushion minor impacts with docks, etc. Many floats also have **spray rails** along the inboard forward portions of the chines. Since water spray is surprisingly destructive to propellers, especially at high r.p.m., these metal flanges are designed to reduce the amount of spray hitting the propeller.

Floats are rated according to the amount of weight they can support, which is based on the weight of the actual volume of fresh water they displace. Fresh water is the standard because sea water is about 3 percent denser than fresh water and can therefore support more weight. If a particular float design displaces 2,500 pounds of fresh water when the float is pushed under the surface, the float can nominally support 2,500

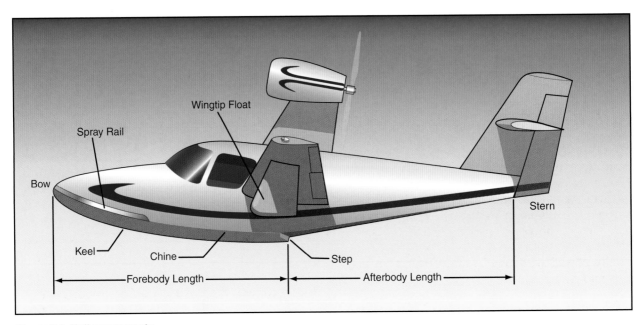

Figure 2-2. Hull components.

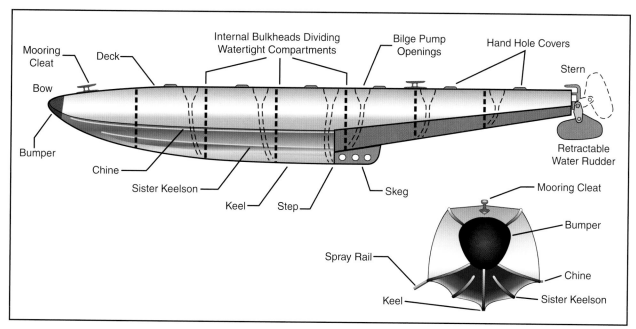

Figure 2-3. Float components.

pounds. A seaplane equipped with two such floats would seemingly be able to support an airplane weighing 5,000 pounds, but the floats would both be completely submerged at that weight. Obviously, such a situation would be impractical, so seaplanes are required to have a buoyancy of 80 percent in excess of that required to support the maximum weight of the seaplane in fresh water. To determine the maximum weight allowed for a seaplane equipped with two floats, divide the total displacement by 180 percent, or 1.8. Using the example of two floats that each displace 2,500 pounds, the total displacement of 5,000 pounds divided by 1.8 gives a maximum weight for the seaplane of 2,778 pounds. Many other considerations determine the suitability of a particular set of floats for a specific type of airplane, and float installations are carefully evaluated by the Federal Aviation Administration (FAA) prior to certification.

All floats are required to have at least four watertight compartments. These prevent the entire float from filling with water if it is ruptured at any point. The floats can support the seaplane with any two compartments flooded, which makes the seaplane difficult to sink.

Most floats have openings with watertight covers along the deck to provide access to the inside of each compartment for inspection and maintenance. There are also smaller holes connected by tubes to the lowest point in each compartment, called the bilge. These bilge pump openings are used for pumping out the bilge water that leaks into the float. The openings are typically closed with small rubber balls that push snugly into place.

Both the lateral and longitudinal lines of a float or hull are designed to achieve a maximum lifting force by

diverting the water and the air downward. The forward bottom portion of a float or hull is designed very much like the bottom of a speedboat. While speedboats are intended to travel at a fairly constant pitch angle, seaplanes need to be able to rotate in pitch to vary the wings' angle of attack and increase lift for takeoffs and landings. The underside of a seaplane float has a sudden break in the longitudinal lines called the **step**. The step provides a means of reducing water drag during takeoff and during high-speed taxi.

At very low speeds, the entire length of the floats supports the weight of the seaplane through buoyancy, that is, the floats displace a weight of water equal to the weight of the seaplane. As speed increases, aerodynamic lift begins to support a certain amount of the weight, and the rest is supported by hydrodynamic lift, the upward force produced by the motion of the floats through the water. Speed increases this hydrodynamic lift, but water drag increases more quickly. To minimize water drag while allowing hydrodynamic lift to do the work of supporting the seaplane on the water, the pilot relaxes elevator back pressure, allowing the seaplane to assume a pitch attitude that brings the aft portions of the floats out of the water. The step makes this possible. When running on the step, a relatively small portion of the float ahead of the step supports the seaplane. Without a step, the flow of water aft along the float would tend to remain attached all the way to the rear of the float, creating unnecessary drag.

The steps are located slightly behind the airplane's center of gravity (CG), approximately at the point where the main wheels are located on a landplane with tricycle gear. If the steps were located too far aft or forward of this point, it would be difficult, if not impossible, to rotate the airplane into a nose-up attitude prior to lifting off. Although steps are necessary, the sharp break along the underside of the float or hull concentrates structural stress into this area, and the disruption in airflow produces considerable drag in flight. The keel under the front portion of each float is intended to bear the weight of the seaplane when it is on dry land. The location of the step near the CG would make it very easy to tip the seaplane back onto the rear of the floats, which are not designed for such loads. The skeg is located behind the step and acts as a sort of chock when the seaplane is on land, making it more difficult to tip the seaplane backward.

Most floatplanes are equipped with retractable **water rudders** at the rear tip of each float. The water rudders are connected by cables and springs to the rudder pedals in the cockpit. While they are very useful in maneuvering on the water surface, they are quite susceptible to damage. The water rudders should be retracted whenever the seaplane is in shallow water or where they might hit objects under the water surface. They are also retracted during takeoff and landing, when dynamic water forces could cause damage.

Seaplane Flight Principles

In the air, seaplanes fly much like landplanes. The additional weight and drag of the floats decrease the airplane's useful load and performance compared to the same airplane with wheels installed. On many airplanes, directional stability is affected to some extent by the installation of floats. This is caused by the length of the floats and the location of their vertical surface area in relation to the airplane's CG. Because the floats present such a large vertical area ahead of the CG, they may tend to increase any yaw or sideslip. To help restore directional stability, an auxiliary fin is often added to the tail. Less aileron pressure is needed to hold the seaplane in a slip. Holding some rudder pressure may be required to maintain coordination in turns, since the cables and springs for the water rudders may tend to prevent the air rudder from streamlining in a turn.

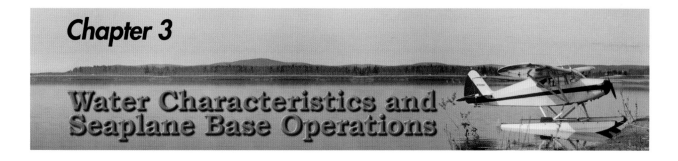
Chapter 3
Water Characteristics and Seaplane Base Operations

CHARACTERISTICS OF WATER

A competent seaplane pilot is knowledgeable in the characteristics of water and how they affect the seaplane. As a fluid, water seeks its own level, and forms a flat, glassy surface if undisturbed. Winds, currents, or objects traveling along its surface create waves and movements that change the surface characteristics.

Just as airplanes encounter resistance in the form of drag as they move through the air, seaplane hulls and floats respond to drag forces as they move through water. Drag varies proportionately to the square of speed. In other words, doubling the seaplane's speed across the water results in four times the drag force.

Forces created when operating an airplane on water are more complex than those created on land. For landplanes, friction acts at specific points where the tires meet the ground. Water forces act along the entire length of a seaplane's floats or hull. These forces vary constantly depending on the pitch attitude, the changing motion of the float or hull, and action of the waves. Because floats are mounted rigidly to the structure of the fuselage, they provide no shock absorbing function, unlike the landing gear of landplanes. While water may seem soft and yielding, damaging forces and shocks can be transmitted directly through the floats and struts to the basic structure of the airplane.

Under calm wind conditions, the smooth water surface presents a uniform appearance from above, somewhat like a mirror. This situation eliminates visual references for the pilot and can be extremely deceptive. If waves are decaying and setting up certain patterns, or if clouds are reflected from the water surface, the resulting distortions can be confusing even for experienced seaplane pilots.

DETERMINING SEA CONDITIONS

The ability to read the water's surface is an integral part of seaplane flying. The interaction of wind and water determine the surface conditions, while tides and currents affect the movement of the water itself. Features along the shore and under the water's surface contribute their effects as well. With a little study, the interplay between these factors becomes clearer.

A few simple terms describe the anatomy and characteristics of waves. The top of a wave is the **crest**, and the low valley between waves is a **trough**. The height of waves is measured from the bottom of the trough to the top of the crest. Naturally, the distance between two wave crests is the wavelength. The time interval between the passage of two successive wave crests at a fixed point is the period of the wave.

Waves are usually caused by wind moving across the surface of the water. As the air pushes the water, ripples form. These ripples become waves in strong or sustained winds; the higher the speed of the wind, or the longer the wind acts on them, the larger the waves. Waves can be caused by other factors, such as underwater earthquakes, volcanic eruptions, or tidal movement, but wind is the primary cause of most waves. [Figure 3-1 on next page]

Calm water begins to show wave motion when the wind reaches about two knots. At this windspeed, patches of ripples begin to form. If the wind stops, surface tension and gravity quickly damp the waves, and the surface returns to its flat, glassy condition. If the wind increases to four knots, the ripples become small waves, which move in the same direction as the wind and persist for some time after the wind stops blowing.

As windspeed increases above four knots, the water surface becomes covered with a complicated pattern of waves. When the wind is increasing, waves become larger and travel faster. If the wind remains at a constant speed, waves develop into a series of evenly spaced parallel crests of the same height.

In simple waves, an object floating on the surface shows that waves are primarily an up and down motion of the water, rather than the water itself moving downwind at the speed of the waves. The floating object describes a circle in the vertical plane, moving upward as the crest approaches, forward and downward as the crest passes, and backward as the trough passes. After each wave passes, the object is at almost the same place as before. The wind does cause floating objects to drift slowly downwind.

While the wind is blowing and adding energy to the water, the resulting waves are commonly referred to as wind waves or **sea**. (Sea is also occasionally used

Terms Used by U.S. Weather Service	Velocity m.p.h.	Estimating Velocities on Land	Estimating Velocities on Sea	
Calm	Less than 1	Smoke rises vertically.	Sea like a mirror.	Check your glassy water technique before water flying under these conditions.
Light Air	1 - 3	Smoke drifts; wind vanes unmoved.	Ripples with the appearance of scales are formed but without foam crests.	
Light Breeze	4 - 7	Wind felt on face; leaves rustle; ordinary vane moves by wind.	Small wavelets, still short but more pronounced; crests have a glassy appearance and do not break.	
Gentle Breeze	8 - 12	Leaves and small twigs in constant motion; wind extends light flag.	Large wavelets; crests begin to break. Foam of glassy appearance. (Perhaps scattered whitecaps.)	Ideal water flying characteristics in protected water.
Moderate Breeze	13 - 18	Dust and loose paper raised; small branches are moved.	Small waves, becoming longer; fairly frequent whitecaps.	
Fresh Breeze	19 - 24	Small trees begin to sway; crested wavelets form in inland water.	Moderate waves; taking a more pronounced long form; many whitecaps are formed. (Chance of some spray.)	This is considered rough water for seaplanes and small amphibians, especially in open water.
Strong Breeze	25-31	Large branches in motion; whistling heard in telegraph wires; umbrellas used with difficulty.	Large waves begin to form; white foam crests are more extensive everywhere. (Probably some spray.)	
Moderate Gale	32-38	Whole trees in motion; inconvenience felt in walking against the wind.	Sea heaps up and white foam from breaking waves begins to be blown in streaks along the direction of the wind.	This type of water condition is for emergency only in small aircraft in inland waters and for the expert pilot.

Figure 3-1. The size of waves is determined by the speed of the wind.

to describe the combined motion of all the factors disturbing the surface.) These waves tend to be a chaotic mix of heights, periods, and wavelengths. Because the wind causes the height to increase faster than the wavelength, they often have relatively steep, pointed crests and rounded troughs. With a windspeed of 12 knots, the waves begin to break at their crests and create foam.

The height of waves depends on three factors: windspeed, length of time the wind blows over the water, and the distance over which the wind acts on the water. As waves move away from the area where they were generated (called a **fetch**), they begin to sort themselves by height and period, becoming regular and evenly spaced. These waves often continue for thousands of miles from where they were generated. **Swell** is the term describing waves that persist outside the fetch or in the absence of the force that generated them. A swell may be large or small, and does not indicate the direction of the wind. The wake of a boat or ship is also a swell.

Unlike wind and current, waves are not deflected much by the rotation of the Earth, but move in the direction in which the generating wind blows. When this wind ceases, water friction and spreading reduce the wave height, but the reduction takes place so slowly that a swell persists until the waves encounter an obstruction, such as a shore. Swell systems from many different directions, even from different parts of the world, may cross each other and interact. Often two or more swell systems are visible on the surface, with a sea wave system developing due to the current wind.

In lakes and sheltered waters, it is often easy to tell wind direction by simply looking at the water's surface. There is usually a strip of calm water along the upwind shore of a lake. Waves are perpendicular to the wind direction. Windspeeds above approximately eight knots leave wind streaks on the water, which are parallel to the wind.

Land masses sculpt and channel the air as it moves over them, changing the wind direction and speed. Wind direction may change dramatically from one part of a lake or bay to another, and may even blow in opposite directions within a surprisingly short distance. Always pay attention to the various wind indicators in the area, especially when setting up for takeoff or landing.

While waves are simply an up and down undulation of the water surface, **currents** are horizontal movements of the water itself, such as the flow of water downstream in a river. Currents also exist in the oceans, where solar heating, the Earth's rotation, and tidal forces cause the ocean water to circulate.

WATER EFFECTS ON OPERATIONS

Compared to operations from typical hard-surface runways, taking off from and landing on water presents several added variables for the pilot to consider. Waves and swell not only create a rough or uneven surface, they also move, and their movement must be considered in addition to the wind direction. Likewise, currents create a situation in which the surface itself is actually moving. The pilot may decide to take off or land with or against the current, depending on the wind, the speed of the current, and the proximity of riverbanks or other obstructions.

While a landplane pilot can rely on windsocks and indicators adjacent to the runway, a seaplane pilot needs to be able to read wind direction and speed from the water itself. On the other hand, the landplane pilot may be restricted to operating in a certain direction because of the orientation of the runway, while the seaplane pilot can usually choose a takeoff or landing direction directly into the wind.

Even relatively small waves and swell can complicate seaplane operations. Takeoffs on rough water can subject the floats to hard pounding as they strike consecutive wave crests. Operating on the surface in rough conditions exposes the seaplane to forces that can potentially cause damage or, in some cases, overturn the seaplane. When a swell is not aligned with the wind, the pilot must weigh the dangers posed by the swell against limited crosswind capability, as well as pilot experience.

On the other hand, calm, glassy water presents a different set of challenges. Since the wind is calm, taxiing and docking are somewhat easier, but takeoffs and landings require special techniques. Takeoff distances may be longer because the wings get no extra lifting help from the wind. The floats seem to adhere more tenaciously to the glassy water surface. When landing, the flat, featureless surface makes it far more difficult to gauge altitude accurately, and reflections can create confusing optical illusions. The specific techniques for glassy water operations are covered in Chapter 4, Seaplane Operations–Preflight and Takeoffs, and Chapter 6, Seaplane Operations–Landing.

Tides are cause for concern when the airplane is beached or moored in shallow water. A rising tide can lift a beached seaplane and allow it to float out to sea if the airplane is not properly secured. Depending on the height of the tide and the topography of the beach, an outgoing tide could leave a beached seaplane stranded far from the water. [Figure 3-2]

Figure 3-2. An outgoing tide can leave a seaplane far from the water. A rising tide can cause a beached seaplane to float away.

Many of the operational differences between land-planes and seaplanes relate to the fact that seaplanes have no brakes. From the time a seaplane casts off, it is usually in continuous motion due to the wind and current, so the pilot must take deliberate action to control this movement. Often these forces can be used to the pilot's advantage to help move the seaplane as desired. Starting the engine, performing the engine runup, and completing most pre-takeoff checks are all accomplished while the seaplane is in motion. The seaplane continues moving after the engine is shut down, and this energy, along with the forces of wind and current, is typically used to coast the seaplane to the desired docking point.

As with land airplanes, the wind tends to make the airplane weathervane, or yaw, until the nose points into the wind. This tendency is usually negligible on landplanes with tricycle landing gear, more pronounced on those with conventional (tailwheel) gear, and very evident in seaplanes. The tendency to weathervane can usually be controlled by using the water rudders while taxiing, but the water rudders are typically retracted prior to takeoff. Weathervaning can create challenges in crosswind takeoffs and landings, as well as in docking or maneuvering in close quarters.

SEAPLANE BASE OPERATIONS

In the United States, rules governing where seaplanes may take off and land are generally left to state and local governments.

Some states and cities are very liberal in the laws regarding the operation of seaplanes on their lakes and waterways, while other states and cities may impose stringent restrictions. The Seaplane Pilots Association publishes the useful Water Landing Directory with information on seaplane facilities, landing areas, waterway use regulations, and local restrictions throughout the United States. Before operating a seaplane on public waters, contact the Parks and Wildlife Department of the state, the State Aeronautics Department, or other authorities to determine the local requirements. In any case, seaplane pilots should always avoid creating a nuisance in any area, particularly in congested marine areas or near swimming or boating facilities.

Established seaplane bases are shown on aeronautical charts and are listed in the Airport/Facility Directory. The facilities at seaplane bases vary greatly, but most include a hard surface ramp for launching, servicing facilities, and an area for mooring or hangaring seaplanes. Many marinas designed for boats also provide seaplane facilities.

Seaplanes often operate in areas with extensive recreational or commercial water traffic. The movements of faster craft, such as speedboats and jet-skis are unpredictable. People towing skiers may be focusing their attention behind the boat and fail to notice a landing seaplane. Swimmers may be nearly invisible, often with just their heads showing among the waves. There is no equivalent of the airport traffic pattern to govern boat traffic, and although right-of-way rules exist on the water, many watercraft operators are unaware of the limits of seaplane maneuverability and may assume that seaplanes will always be able to maneuver to avoid them. Many times, the seaplane itself is an object of curiosity, drawing water traffic in the form of interested onlookers.

When seaplane operations are conducted in bush country, regular or emergency facilities are often limited or nonexistent. The terrain and waterways are frequently hazardous, and any servicing becomes the individual pilot's responsibility. Prior to operating in an unfamiliar area away from established seaplane facilities, obtain the advice of FAA Accident Prevention Counselors or experienced seaplane pilots who are familiar with the area.

Chapter 4

Seaplane Operations - Preflight and Takeoffs

PREFLIGHT INSPECTION

Begin the preflight inspection with a thorough review of the existing local weather, destination weather, and water conditions. This weather evaluation should include the direction and speed of the wind to determine their effects on takeoffs, landings, and other water operations.

The preflight inspection of a seaplane is somewhat different from that of a landplane. Inspecting a seaplane on the water is complicated by the need to reposition the seaplane to gain access to all parts of the airframe. On the other hand, preflighting a seaplane on land may create certain challenges because the wings and tail surfaces may be out of reach and difficult to inspect when standing on the ground.

The following preflight description omits many items that are identical in landplanes and seaplanes in order to emphasize the differences between the two procedures. The process and the equipment to be checked vary from airplane to airplane, but the following description provides a general idea of the preflight inspection for a typical high wing, single-engine floatplane. As always, follow the procedures recommended in the Airplane Flight Manual (AFM) or Pilot's Operating Handbook (POH).

If the seaplane is in the water during the preflight, take a good look at how it sits on the surface. This can provide vital clues to the presence of water in the floats, as well as to the position of the center of gravity. Is the seaplane lower in the water than it should be, given its load? Is one wing lower than the other, or is one float riding noticeably lower in the water than the other? Are the sterns of the floats low in the water? If any of these signs are present, suspect a flooded float compartment or an improperly loaded seaplane. At more than 8 pounds per gallon, even a relatively small amount of water in a float compartment can seriously affect both useful load and center of gravity (CG).

In the cockpit, verify that the throttle is closed, the mixture control is full lean, and the magnetos and master switch are turned off. Lower the water rudders and check for any stiffness or binding in the action of the cables. Check that necessary marine and safety equipment, such as life vests, lines (ropes), anchors, and paddles are present, in good condition, and stowed correctly. Obtain the bilge pump and fuel sample cup.

Standing on the front of the float, inspect the propeller, forward fuselage, and wing. Check the usual items, working from the nose toward the tail. Water spray damage to the propeller looks similar to gravel damage, and must be corrected by a mechanic. Check the oil and fuel levels and sample the fuel, ensuring that it is the proper grade and free of contaminants. Naturally, the most likely contaminant in seaplane fuel tanks is water. Pay extra attention to the lubrication of all hinges. Not only does lubrication make movement easier, but a good coating of the proper lubricant keeps water out and prevents corrosion. Look for any blistering or bubbling of the paint, which may indicate corrosion of the metal underneath. Check the security of the float struts and their attachment fittings. Be careful moving along the float, and pay attention to wing struts, mooring lines, and other obstacles. If the seaplane is on land, do not stand on the floats aft of the step or the seaplane may tip back.

Next, inspect the float itself. Water forces can create very high loads and lead to cumulative damage. Look carefully for signs of stress, such as distortion or buckling of the skin, dents, or loose rivets. The chines should form a continuous smooth curve from front to back, and there should be no bends or kinks along the flange. If the floats are made of fiberglass or composite materials, look carefully for surface cracks, abrasions, or signs of delamination. Check the spreader bars between the floats, and look at the bracing wires and their fittings. Any sign of movement, loose fasteners, broken welds, or a bracing wire that is noticeably tighter or looser than the others is cause for concern. Check for signs of corrosion, especially if the seaplane has been operated in salt water. Although corrosion is less of an issue with composite floats, be sure to check metal fittings and fasteners. [Figure 4-1]

Figure 4-1. A preflight inspection with the seaplane on land provides an opportunity to thoroughly examine the floats below the waterline. Note the spray rail on the inboard chine of the far float in this photo.

Use the bilge pump to remove any accumulated water from each watertight compartment. The high dynamic water pressure and the physical stresses of takeoffs and landings can momentarily open tiny gaps between float components, allowing small amounts of water to enter. Conversely, sitting idle in the water also results in a small amount of seepage and condensation. While it is normal to pump a modest amount of water from each compartment, more than a quart or so may indicate a problem that should be checked by a qualified aircraft mechanic experienced in working on floats. Normal is a relative term, and experience will indicate how much water is too much. [Figure 4-2]

Figure 4-2. Bilge pump openings are closed with a soft rubber ball.

If pumping does not remove any water from a compartment, the tube running from the bilge pump opening to the bottom of the compartment may be damaged or loose. If this is the case, there could be a significant amount of water in the compartment, but the pump is unable to pull it up. [Figure 4-3] Be sure to replace the plugs firmly in each bilge pump opening.

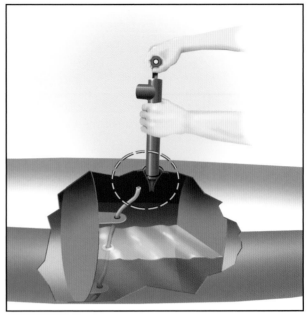

Figure 4-3. Be suspicious if pumping does not remove a small amount of water. If the bilge pump tube is damaged, there may be water in the compartment that the pump cannot remove.

At the stern of the float, check the aft bulkhead, or transom. This area is susceptible to damage from the water rudder moving beyond its normal range of travel. Carefully check the skin for any pinholes or signs of damage from contact with the water rudder or hinge hardware. Inspect the water rudder retraction and steering mechanism and look over the water rudder for any damage. Remove any water weeds or other debris lodged in the water rudder assembly. Check the water rudder cables that run from the float to the fuselage. [Figure 4-4]

Figure 4-4. Inspect the water rudders, cables, springs, and pulleys for proper operation.

To check the empennage area, untie the seaplane, gently push it away from the dock, and turn it 90° so the tail extends over the dock. Take care not to let the water rudders contact the dock. In addition to the normal empennage inspection, check the cables that connect the water rudders to the air rudder. With the air rudder centered, look at the back of the floats to see that the water rudders are also centered. (On some systems, retracting the water rudders disengages them from the air rudder.) If the seaplane has a ventral fin to improve directional stability, this is the time to check it. Spray frequently douses the rear portion of the seaplane, so be particularly alert for signs of corrosion in this area.

With the empennage inspection complete, continue turning the seaplane to bring the other float against the dock, and tie it to the dock. Inspect the fuselage, wing, and float on this side. If the seaplane has a door on only one side, turn the seaplane so the door is adjacent to the dock when the inspection is complete.

When air temperatures drop toward freezing, ice becomes a matter for concern. Inspect the float compartments and water rudders for ice, and consider the possibility of airframe icing during takeoff due to freezing spray. Water expands as it freezes, and this expansion can cause serious damage to floats. A large amount of water expanding inside a float could cause seams to burst, but even a tiny amount of water freezing and expanding inside a seam can cause severe leakage problems. Many operators who remove their floats for the winter store them upside down with the compartment covers off to allow thorough drainage. When the time comes to reinstall the floats, it's a good idea to look for any bugs or small animals that might have made a home in the floats.

STARTING THE ENGINE

Compared to a landplane, a seaplane's starting procedures are somewhat different. Before starting the engine, the seaplane usually needs to be pushed away from the dock, and quite often, it is the pilot who pushes off. Therefore, the pilot should perform as many of the items on the starting checklist as possible prior to shoving off. This includes briefing passengers and seeing that they have fastened their seatbelts. The passenger briefing should include procedures for evacuation, the use of flotation gear, and the location and operation of regular and emergency exits. All passengers are required to be familiar with the operation of seatbelts and shoulder harnesses (if installed). When the engine is primed and ready to start, the pilot leaves the cockpit, shoves off, returns to the pilot's seat, quickly turns on the master switch and magnetos, verifies that the propeller area is clear, and starts the engine. With oil pressure checked, idle r.p.m. set, and the seaplane taxiing in the desired direction, the pilot then fastens the seatbelt and shoulder harness, secures the door, and continues preparing for takeoff.

When a qualified person is available to help launch the seaplane, the pilot can strap in, close the door, and start the engine while the helper holds the seaplane. In most situations, the helper should position the seaplane so it is facing outward, perpendicular to the dock. It is very important that the helper have experience in the proper handling of seaplanes, otherwise an innocent mistake could cause serious damage to the seaplane or to nearby boats, structures, or other seaplanes.

TAXIING AND SAILING

One major difference between taxiing a landplane and taxiing a seaplane is that the seaplane is virtually always in motion, and there are no brakes. When idling, a landplane usually remains motionless, and when moving, brakes can be used to control its speed or bring it to a stop. But once untied, the seaplane floats freely along the water surface and constantly moves due to the forces of wind, water currents, propeller thrust, and inertia. It is important that the seaplane pilot be familiar with the existing wind and water conditions, plan an effective course of action, and mentally stay ahead of the seaplane.

There are three basic positions or attitudes used in moving a seaplane on the water, differentiated by the position of the floats and the speed of the seaplane through the water. They are the idling or displacement position, the plowing position, and the planing or step position.

IDLING POSITION

In the **idling position** or **displacement position**, the buoyancy of the floats supports the entire weight of the seaplane and it remains in an attitude similar to being at rest on the water. Engine r.p.m. is kept as low as possible to control speed, to keep the engine from overheating, and to minimize spray. In almost all circumstances, the elevator control should be held all the way back to keep the nose as high as possible and minimize spray damage to the propeller. This also improves maneuverability by keeping more of the water rudder underwater. The exception is when a strong tailwind component or heavy swells could allow the wind to lift the tail and possibly flip the seaplane over. In such conditions, hold the elevator control forward enough to keep the tail down. [Figure 4-5 on next page]

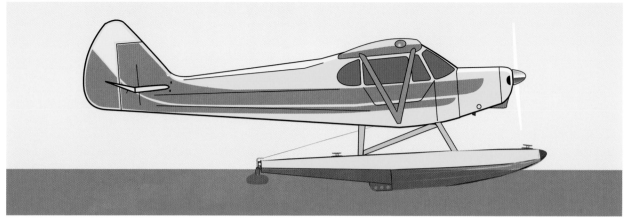

Figure 4-5. Idling position. The engine is at idle r.p.m., the seaplane moves slowly, the attitude is nearly level, and buoyancy supports the seaplane.

Use the idling or displacement position for most taxiing operations, and keep speeds below 6-7 knots to minimize spray getting to the propeller. It is especially important to taxi at low speed in congested or confined areas because inertia forces at higher speeds allow the seaplane to coast farther and serious damage can result from even minor collisions. Cross boat wakes or swells at a 45° angle, if possible, to minimize pitching or rolling and the possibility of an upset.

PLOWING POSITION

Applying power causes the center of buoyancy to shift back, due to increased hydrodynamic pressure on the bottoms of the floats. This places more of the seaplane's weight behind the step, and because the floats are narrower toward the rear, the sterns sink farther into the water. Holding the elevator full up also helps push the tail down due to the increased airflow from the propeller. The **plowing position** creates high drag, requiring a relatively large amount of power for a modest gain in speed. Because of the higher r.p.m., the propeller may pick up spray even though the nose is high. The higher engine power combined with low cooling airflow creates a danger of heat buildup in the engine. Monitor engine temperature carefully to avoid overheating. Taxiing in the plowing position is not

recommended. It is usually just the transitional phase between idle taxi and planing. [Figure 4-6]

PLANING OR STEP POSITION

In the **planing position**, most of the seaplane's weight is supported by hydrodynamic lift rather than the buoyancy of the floats. (Because of the wing's speed through the air, aerodynamic lift may also be supporting some of the weight of the seaplane.) Hydrodynamic lift depends on movement through the water, like a water ski. As the float moves faster through the water, it becomes possible to change the pitch attitude to raise the rear portions of the floats clear of the water. This greatly reduces water drag, allowing the seaplane to accelerate to lift-off speed. This position is most often called **on the step**. [Figure 4-7]

There is one pitch attitude that produces the minimum amount of drag when the seaplane is on the step. An experienced seaplane pilot can easily find this "sweet spot" or "slick spot" by the feel of the floats on the water, but the beginning seaplane pilot usually needs to rely on gauging the position of the nose on the horizon. If the nose is considerably high, the rear portions of the floats contact the water, drag increases, and the

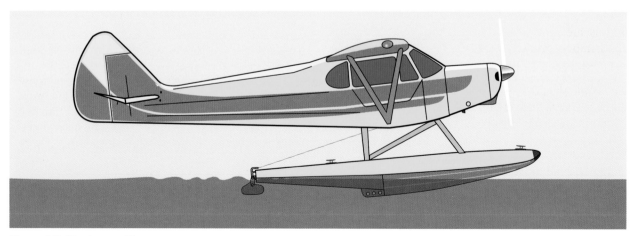

Figure 4-6. Plowing position.

4-4

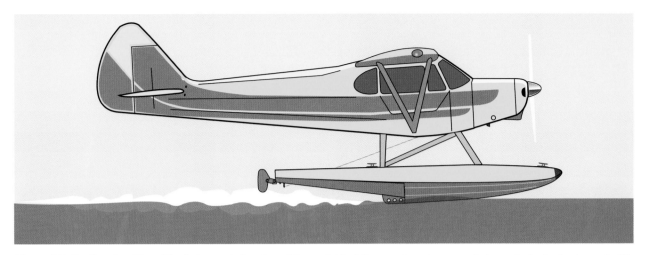

Figure 4-7. On the step. The attitude is nearly level, and the weight of the seaplane is supported mostly by hydrodynamic lift. Behind the step, the floats are essentially clear of the water.

seaplane tends to start settling back into more of a plowing position. If the nose is held only slightly higher than the ideal planing attitude, the seaplane may remain on the step but take much longer to accelerate to rotation speed. On the other hand, if the nose is too low, more of the front portion of the float contacts the water, creating more drag. This condition is called dragging, and as the nose pulls down and the seaplane begins to slow, it can sometimes feel similar to applying the brakes in a landplane.

To continue to taxi on the step instead of taking off, reduce the power as the seaplane is eased over onto the step. More power is required to taxi with a heavy load. However, 65 to 70 percent of maximum power is a good starting point.

Taxiing on the step is a useful technique for covering long distances on the water. Carefully reducing power as the seaplane comes onto the step stops acceleration so that the seaplane maintains a high speed across the water, but remains well below flying speed. At these speeds, the water rudders must be retracted to prevent damage, but there is plenty of airflow for the air rudder. With the seaplane on the step, gentle turns can be made by using the air rudder and the ailerons, always maintaining a precise planing attitude with elevator. The ailerons are positioned into the turn, except when aileron into the wind is needed to keep the upwind wing from lifting.

Step taxiing should only be attempted in areas where the pilot is confident there is sufficient water depth, no floating debris, no hidden obstructions, and no other water traffic nearby. It can be difficult to spot floating hazards at high speeds, and an encounter with a floating log or other obstruction could tear open a float. Your seaplane is not as maneuverable as craft that were designed for the water, so avoiding other vessels is much more difficult. Besides the obvious danger of collision, other water traffic creates dangerous wakes, which are a

much more frequent cause of damage. If you see that you are going to cross a wake, reduce power to idle and idle taxi across it, preferably at an angle. Never try to step taxi in shallow water. If the floats touch bottom at high speed, the sudden drag is likely to flip the seaplane.

From either the plowing or the step position, when power is reduced to idle, the seaplane decelerates quite rapidly and eventually assumes the displacement or idle position. Be careful to use proper flight control pressures during the deceleration phase because as weight is transferred toward the front of the floats and drag increases, some seaplanes have a tendency to nose over. Control this with proper use of the elevator.

TURNS
At low speeds and in light winds, make turns using the water rudders, which move in conjunction with the air rudder. As with a landplane, the ailerons should be positioned to minimize the possibility of the wind lifting a wing. In most airplanes, left turns are somewhat easier and can be made tighter than right turns because of torque. If water rudders have the proper amount of movement, most seaplanes can be turned within a radius less than the span of the wing in calm conditions or a light breeze. Water rudders are usually more effective at slow speeds because they are acting in comparatively undisturbed water. At higher speeds, the stern of the float churns the adjacent water, causing the water rudder to become less effective. The dynamic pressure of the water at high speeds may tend to force the water rudders to swing up or retract, and the pounding can cause damage. For these reasons, water rudders should be retracted whenever the seaplane is moving at high speed.

The weathervaning tendency is more evident in seaplanes, and the taxiing seaplane pilot must be constantly aware of the wind's effect on the ability to maneuver. In stronger winds, weathervaning forces may make it difficult to turn downwind. Often a short burst of power provides sufficient air over the rudder to overcome weathervaning. Since the elevator is held all the way up, the airflow also forces the tail down, making the water rudders more effective. Short bursts of power are preferable to a longer, continuous power application. With continuous power, the seaplane accelerates, increasing the turn radius. The churning of the water in the wake of the floats also makes the water rudders less effective. At the same time, low cooling airflow may cause the engine to heat up.

During a high speed taxiing turn, centrifugal force tends to tip the seaplane toward the outside of the turn. When turning from an upwind heading to a downwind heading, the wind force acts in opposition to centrifugal force, helping stabilize the seaplane. On the other hand, when turning from downwind to upwind, the wind force against the fuselage and the underside of the wing increases the tendency for the seaplane to lean to the outside of the turn, forcing the downwind float deeper into the water. In a tight turn or in strong winds, the combination of these two forces may be sufficient to tip the seaplane to the extent that the downwind float submerges or the outside wing drags in the water, and may even flip the seaplane onto its back. The further

the seaplane tips, the greater the effect of the crosswind, as the wing presents more vertical area to the wind force. [Figure 4-8]

When making a turn into the wind from a crosswind condition, often all that is necessary to complete the turn is to neutralize the air rudder and allow the seaplane to weathervane into the wind. If taxiing directly downwind, use the air rudder momentarily to get the turn started, then let the wind complete the turn. Sometimes opposite rudder may be needed to control the rate of turn.

Stronger winds may make turns from upwind to downwind more difficult. The plow turn is one technique for turning downwind when other methods are inadequate, but this maneuver is only effective in certain seaplanes. It takes advantage of the same factor that reduces a floatplane's yaw stability in flight: the large vertical area of the floats forward of the center of gravity. In the plowing attitude, the front portion of each float comes out of the water, presenting a large vertical surface for the wind to act upon. This tends to neutralize the weathervaning force, allowing the turn to proceed. At the same time, the center of buoyancy shifts back. Since this is the axis around which the seaplane pivots while

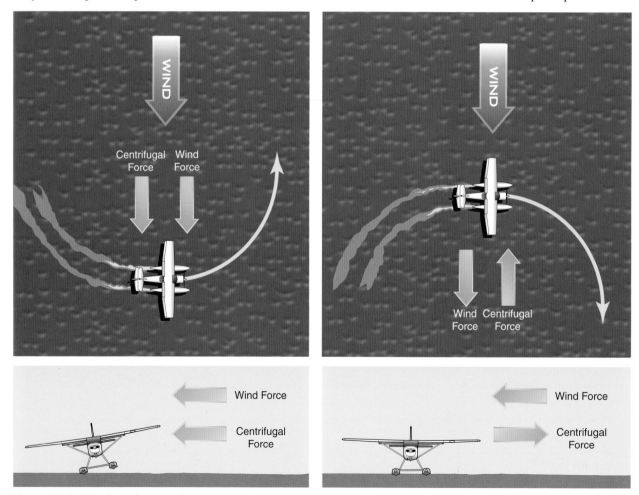

Figure 4-8. Wind effects in turns. When the wind and centrifugal force act in the same direction, the downwind float can be forced underwater. When the wind is countered by centrifugal force, the seaplane is more stable.

on the water, more of the fuselage is now forward of the axis and less is behind, further decreasing the weathervaning tendency. In some seaplanes, this change is so pronounced in the plowing attitude that they experience reverse weathervaning, and tend to turn downwind rather than into the wind. Experienced seaplane pilots can sometimes use the throttle as a turning device in high wind conditions by increasing power to cause a nose-up position when turning downwind, and decreasing power to allow the seaplane to weathervane into the wind. [Figure 4-9]

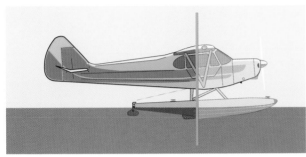
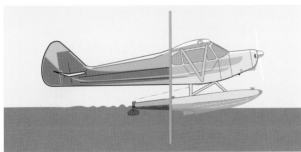

Figure 4-9. In the plowing position, the exposed area at the front of the floats, combined with the rearward shift of the center of buoyancy, can help to counteract the weathervaning tendency.

To execute a plow turn, begin with a turn to the right, then use the weathervaning force combined with full left rudder to turn back to the left. As the seaplane passes its original upwind heading, add enough power to place it into the plow position, continuing the turn with the rudder. As the seaplane comes to the downwind heading, reduce power and return to an idle taxi. From above, the path of the turn looks like a question mark. [Figure 4-10]

Plow turns are useful only in very limited situations because they expose the pilot to a number of potential dangers. They should not be attempted in rough water or gusty conditions. Floatplanes are least stable when in the plowing attitude, and are very susceptible to capsizing. In spite of the nose-high attitude, the high power setting often results in spray damage to the propeller. In most windy situations, it is much safer to sail the seaplane backward (as explained in the next section) rather than attempt a plow turn.

When the seaplane is on the step, turns involve careful balancing of several competing forces. As the rate of

turn increases, the floats are being forced to move somewhat sideways through the water, and they resist this sideways motion with drag, much like an airplane fuselage in a skidding turn. More power is required to overcome this drag and maintain planing speed. This skidding force also tends to roll the seaplane toward the outside of the turn, driving the outside float deeper into the water and adding more drag on that side. To prevent this, use aileron into the turn to keep the outside wing from dropping. Once full aileron into the step turn is applied, any further roll to the outside can only be stopped by reducing the rate of turn, so pay careful attention to the angle of the wings and the feel of the water drag on the floats to catch any indication that the outside float is starting to submerge. When stopping a step turn, always return to a straight path before reducing power.

At step taxi speeds, the centrifugal force in a turn is far greater than at idle taxi speed, so the forces involved in turning from downwind to upwind are proportionately more dangerous, especially in strong winds. Chances are, by the time a pilot discovers that the outside float is going under, the accident is almost inevitable. However, immediate full rudder out of the turn and power reduction may save the situation by reversing

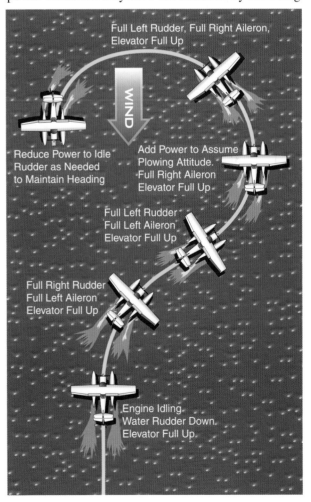

Figure 4-10. Plow turn from upwind to downwind.

the centrifugal force and allowing the buried float to come up.

SAILING

Landplane pilots are accustomed to taxiing by pointing the nose of the airplane in the desired direction and rolling forward. In seaplane operations, there are often occasions when it is easier and safer to move the seaplane backward or to one side because wind, water conditions, or limited space make it impractical to attempt a turn. If there is a significant wind, a seaplane can be guided into a space that might seem extremely cramped to an inexperienced pilot. **Sailing** is a method of guiding the seaplane on the water using the wind as the main motive force. It is a useful technique for maneuvering in situations where conventional taxiing is undesirable or impossible. Since the seaplane automatically aligns itself so the nose points into the wind, sailing in a seaplane usually means moving backward.

In light wind conditions with the engine idling or off, a seaplane naturally weathervanes into the wind. If the pilot uses the air rudder to swing the tail a few degrees, the seaplane sails backward in the direction the tail is pointed. This is due to the keel effect of the floats, which tends to push the seaplane in the direction the sterns of the floats are pointing. In this situation, lift the water rudders, since their action is counter to what is desired. When sailing like this, the sterns of the floats have become the front, as far as the water is concerned, but the rear portions of the floats are smaller and therefore not as buoyant. If the wind is strong and speed starts to build up, the sterns of the floats could start to

submerge and dig into the water. Combined with the lifting force of the wind over the wings, the seaplane could conceivably flip over backward, so use full forward elevator to keep the sterns of the floats up and the seaplane's nose down. Adding power can also help keep the floats from submerging.

If enough engine power is used to exactly cancel the backward motion caused by the wind, the seaplane is not moving relative to the water, so keel effect disappears. However, turning the fuselage a few degrees left or right provides a surface for the wind to push against, so the wind will drive the seaplane sideways in the direction the nose is pointed. Combining these techniques, a skilled pilot can sail a seaplane around obstacles and into confined docking spaces. [Figure 4-11]

Figure 4-12 shows how to position the controls for the desired direction of motion in light or strong winds. With the engine off, lowering the wing flaps and opening the cabin doors increases the air resistance and thus adds to the effect of the wind. This increases sailing speed but may reduce the effect of the air rudder. If sailing with the engine off results in too much motion downwind, but an idling engine produces too much thrust, adding carburetor heat or turning off one magneto can reduce the engine power slightly. Avoid using carburetor heat or running on one magneto for extended periods. Instead, start the engine briefly to slow down.

Where currents are a factor, such as in strong tidal flows or a fast flowing river, sailing techniques must

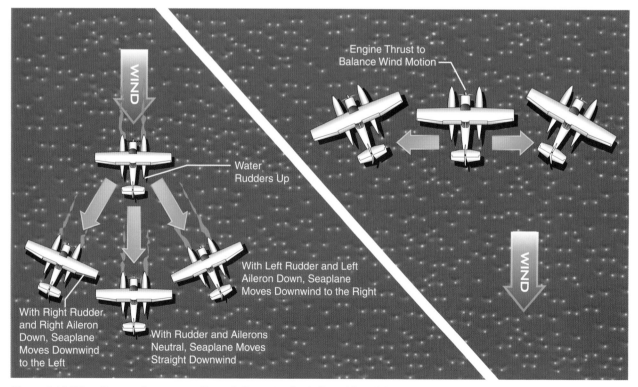

Figure 4-11. When the seaplane moves through the water, keel effect drives it in the direction the tail is pointed. With no motion through the water, the wind pressure on the fuselage pushes the seaplane toward the side the nose is pointed.

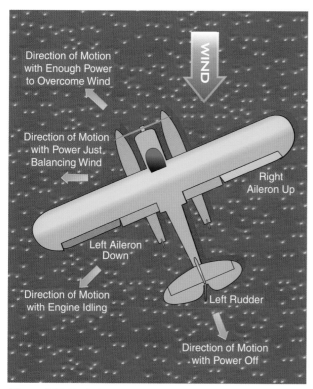

Figure 4-12. By balancing wind force and engine thrust, it is possible to sail sideways or diagonally forward. Of course, reversing the control positions from those illustrated permits the pilot to sail to the opposite side.

incorporate the movement of the water along with the wind. The current may be a help or a hindrance, or change from a help to a hindrance when the pilot attempts to change direction. The keel effect only works when the floats are moving through the water. If the current is moving the seaplane, there may be little or no motion relative to the water, even though the seaplane is moving relative to the shore. Using wind, current, and thrust to track the desired course requires careful planning and a thorough understanding of the various forces at work.

With the engine shut down, most flying boats sail backward and toward whichever side the nose is pointed, regardless of wind velocity, because the hull does not provide as much keel effect as floats in proportion to the side area of the seaplane above the waterline. To sail directly backward in a flying boat, release the controls and let the wind steer. Sailing is an essential part of seaplane operation. Since each type of seaplane has its own peculiarities, practice

sailing until thoroughly familiar with that particular type. Practice in large bodies of water such as lakes or bays, but sufficiently close to a prominent object in order to evaluate performance.

Before taxiing into a confined area, carefully evaluate the effects of the wind and current, otherwise the seaplane may be driven into obstructions. With a seaplane of average size and power at idle, a water current of 5 knots can offset a wind velocity of 25 knots in the opposite direction. This means that a 5 knot current will carry the seaplane against a 25 knot wind. Differential power can be used to aid steering in multi-engine seaplanes.

PORPOISING

Porpoising is a rhythmic pitching motion caused by dynamic instability in forces along the float bottoms while on the step. An incorrect planing attitude sets off a cyclic oscillation that steadily increases in amplitude unless the proper pitch attitude is reestablished. [Figure 4-13]

A seaplane travels smoothly across the water on the step only if the floats or hull remain within a moderately tolerant range of pitch angles. If the nose is held too low during planing, water pressure in the form of a small crest or wall builds up under the bows of the floats. Eventually, the crest becomes large enough that the fronts of the floats ride up over the crest, pitching the bows upward. As the step passes over the crest, the floats tip forward abruptly, digging the bows a little deeper into the water. This builds a new crest in front of the floats, resulting in another oscillation. Each oscillation becomes increasingly severe, and if not corrected, will cause the seaplane to nose into the water, resulting in extensive damage or possible capsizing. A second type of porpoising can occur if the nose is held too high while on the step. Porpoising can also cause a premature lift-off with an extremely high angle of attack, which can result in a stall and a subsequent nose-down drop into the water. Porpoising occurs during the takeoff run if the planing angle is not properly controlled with elevator pressure just after passing through the "hump" speed. The pitching created when the seaplane encounters a swell system while on the step can also initiate porpoising. Usually, porpoising does not start until the seaplane has passed a degree or two beyond the acceptable planing angle range, and

Figure 4-13. Porpoising increases in amplitude if not corrected promptly.

does not cease until after the seaplane has passed out of the critical range by a degree or two.

If porpoising occurs due to a nose-low planing attitude, stop it by applying timely back pressure on the elevator control to prevent the bows of the floats from digging into the water. The back pressure must be applied and maintained until porpoising stops. If porpoising does not stop by the time the second oscillation occurs, reduce the power to idle and hold the elevator control back firmly so the seaplane settles onto the water with no further instability. Never try to "chase" the oscillations, as this usually makes them worse and results in an accident.

Pilots must learn and practice the correct pitch attitudes for takeoff, planing, and landing for each type of seaplane until there is no doubt as to the proper angles for the various maneuvers. The upper and lower limits of these pitch angles are established by the design of the seaplane; however, changing the seaplane's gross weight, wing flap position, or center of gravity location also changes these limits. Increased weight increases the displacement of the floats or hull and raises the lower limit considerably. Extending the wing flaps frequently trims the seaplane to the lower limit at lower speeds, and may lower the upper limit at high speeds. A forward center of gravity increases the possibility of high angle porpoising, especially during landing.

SKIPPING

Skipping is a form of instability that may occur when landing at excessive speed with the nose at too high a pitch angle. This nose-up attitude places the seaplane at the upper trim limit of stability and causes the seaplane to enter a cyclic oscillation when touching the water, which results in the seaplane skipping across the surface. This action is similar to skipping flat stones across the water. Skipping can also occur by crossing a boat wake while taxiing on the step or during a takeoff. Sometimes the new seaplane pilot confuses a skip with a porpoise, but the pilot's body sensations can quickly distinguish between the two. A skip gives the body vertical "G" forces, similar to bouncing a landplane. Porpoising is a rocking chair type forward and aft motion feeling.

To correct for skipping, first increase back pressure on the elevator control and add sufficient power to prevent the floats from contacting the water. Then establish the proper pitch attitude and reduce the power gradually to allow the seaplane to settle gently onto the water. Skipping oscillations do not tend to increase in amplitude, as in porpoising, but they do subject the floats and airframe to unnecessary pounding and can lead to porpoising.

TAKEOFFS

A seaplane takeoff may be divided into four distinct phases: (1) The displacement phase, (2) the hump or plowing phase, (3) the planing or on the step phase, and (4) the lift-off.

The displacement phase should be familiar from the taxiing discussion. During idle taxi, the displacement of water supports nearly all of the seaplane's weight. The weight of the seaplane forces the floats down into the water until a volume that weighs exactly as much as the seaplane has been displaced. The surface area of the float below the waterline is called the wetted area, and it varies depending on the seaplane's weight. An empty seaplane has less wetted area than when it is fully loaded. Wetted area is a major factor in the creation of drag as the seaplane moves through the water.

As power is applied, the floats move faster through the water. The water resists this motion, creating drag. The forward portion of the float is shaped to transform the horizontal movement through the water into an upward lifting force by diverting the water downward. Newton's Third Law of Motion states that for every action, there is an equal and opposite reaction, and in this case, pushing water downward results in an upward force known as hydrodynamic lift.

In the plowing phase, hydrodynamic lift begins pushing up the front of the floats, raising the seaplane's nose and moving the center of buoyancy aft. This, combined with the downward pressure on the tail generated by holding the elevator control all the way back, forces the rear part of the floats deeper into the water. This creates more wetted area and consequently more drag, and explains why the seaplane accelerates so slowly during this part of the takeoff.

This resistance typically reaches its peak just before the floats are placed into a planing attitude. Figure 4-14 shows a graph of the drag forces at work during a seaplane takeoff run. The area of greatest resistance is referred to as the hump because of the shape of the water drag curve. During the plowing phase, the increasing water speed generates more and more hydrodynamic lift. With more of the weight supported by hydrodynamic lift, proportionately less is supported by displacement and the floats are able to rise in the water. As they do, there is less wetted area to cause drag, which allows more acceleration, which in turn increases hydrodynamic lift. There is a limit to how far this cycle can go, however, because as speed builds, so does the amount of drag on the remaining wetted area. Drag increases as the square of speed, and eventually drag forces would balance the power output of the engine and the seaplane would continue along the surface without further acceleration.

Seaplanes have been built with sufficient power to accelerate to takeoff speed this way, but fortunately the step was invented, and it makes further acceleration possible without additional power. After passing over the hump, the seaplane is traveling fast enough that its weight can be supported entirely by hydrodynamic lift. Relaxing the back pressure on the elevator control allows the float to rock up onto the step, and lifts the

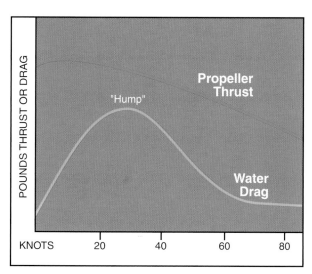

Figure 4-14. This graph shows water drag and propeller thrust during a takeoff run.

rear portions of the floats clear of the water. This eliminates all of the wetted area aft of the step, along with the associated drag.

As further acceleration takes place, the flight controls become more responsive, just as in a landplane. Elevator deflection is gradually reduced to hold the required planing attitude. As the seaplane continues to accelerate, more and more weight is being supported by the aerodynamic lift of the wings and water resistance continues to decrease. When all of the weight is transferred to the wings, the seaplane becomes airborne.

Several factors greatly increase the water drag or resistance, such as heavy loading of the seaplane or glassy water conditions. In extreme cases, the drag may exceed the available thrust and prevent the seaplane from becoming airborne. This is particularly true when operating in areas with high density altitudes (high elevations/high temperatures) where the engine cannot develop full rated power. For this reason the pilot should practice takeoffs using only partial power to simulate the longer takeoff runs needed when operating where the density altitude is high and/or the seaplane is heavily loaded. This practice should be conducted under the supervision of an experienced seaplane instructor, and in accordance with any cautions or limitations in the AFM/POH. Plan for the additional takeoff area required, as well as the flatter angle of climb after takeoff, and allow plenty of room for error.

Use all of the available cues to verify the wind direction. Besides reading the water, pick up clues to the wind's direction from wind indicators and streamers on the masts of moored boats, flags on flagpoles, or rising smoke. A boat moored to a buoy points into the wind, but be aware that it may have a stern anchor as well, preventing it from pointing into the wind.

Waterfowl almost always align themselves facing into the wind.

Naturally, be sure you have enough room for takeoff. The landing distance of a seaplane is much shorter than that required for takeoff, and many pilots have landed in areas that have turned out to be too short for takeoff. If you suspect that the available distance may be inadequate, consider reducing weight by leaving some of your load behind or wait for more favorable weather conditions. A takeoff that would be dangerous on a hot, still afternoon might be accomplished safely on the following morning, with cooler temperatures and a brisk wind.

In addition to wind, consider the effects of the current when choosing the direction for takeoff. Keep in mind that when taxiing in the same direction as the current, directional control may be reduced because the seaplane is not moving as quickly through the water. In rivers or tidal flows, make crosswind or calm wind takeoffs in the same direction as the current. This reduces the water forces on the floats. Suppose the seaplane lifts off at 50 knots and the current is 3 knots. If winds are calm, the seaplane needs a water speed of 47 knots to take off downstream, but must accelerate to a water speed of 53 knots to become airborne against the current. This difference of 6 knots requires a longer time on the water and generates more stress on the floats. The situation becomes more complex when wind is a factor. If the wind is blowing against the current, its speed can help the wings develop lift sooner, but will raise higher waves on the surface. If the wind is in the same direction as the current, at what point does the speed of the wind make it more worthwhile to take off against the current? In the previous example, a wind velocity of 3 knots would exactly cancel the benefit of the current, since the air and water would be moving at the same speed. In most situations, take off into the wind if the speed of the wind is greater than the current.

Unlike landplane operations at airports, many other activities are permitted in waters where seaplane operations are conducted. Seaplane pilots encounter a variety of objects on the water, some of which are nearly submerged and difficult to see. These include items that are stationary, such as pilings and buoys, and those that are mobile, like logs, swimmers, water skiers, and a variety of watercraft. Before beginning the takeoff, it is a good practice to taxi along the intended takeoff path to check for any hazardous objects or obstructions.

Make absolutely sure the takeoff path ahead is free of boats, swimmers, and other water traffic, and be sure it will remain so for the duration of the takeoff run. Powerboats, wind-surfers, and jet-skis can move quickly and change direction abruptly. As the seaplane's nose comes up with the application of full power, the view ahead may be completely blocked by the cowling. Check to the sides and behind the seaplane as well as straight ahead, since many watercraft move much faster than the normal taxi speed and may be passing the seaplane from behind. In addition to the vessels themselves, also scan for their wakes and try to anticipate where the wakes will be during takeoff. Operators of motorboats and other watercraft often do not realize the hazard caused by moving their vessels across the takeoff path of a seaplane. It is usually better to delay takeoff and wait for the swells to pass rather than encountering them at high speed. Even small swells can cause dangerous pitching or rolling for a seaplane, so taxi across them at an angle rather than head-on. Remember to check for other air traffic and make any appropriate radio calls.

Be sure to use the pre-takeoff checklist on every takeoff. All checks are performed as the seaplane taxies, including the engine runup. Hold the elevator control all the way back throughout the runup to minimize spray around the propeller. If there is significant wind, let the seaplane turn into the wind for the runup. As r.p.m. increases, the nose rises into the plowing position and the seaplane begins to accelerate. Since this is a relatively unstable position, performing the runup into the wind minimizes the possibility of crosswinds, rough water, or gusts upsetting the seaplane. Waste no time during the runup checks, but be thorough and precise. Taxi speed will drop as soon as the power is reduced.

Water rudders are normally retracted before applying takeoff power. The buffeting and dynamic water pressure during a takeoff can cause serious damage if the water rudders are left down.

As full power is applied during takeoff in most seaplanes, torque and P-factor tend to force the left float down into the water. Right rudder pressure helps to maintain a straight takeoff path. In some cases, left aileron may also help to counter the tendency to turn left at low speeds, by increasing drag on the right side of the seaplane.

Density altitude is particularly important in seaplane flying. High, hot, and humid conditions reduce engine power and propeller efficiency, and the seaplane must also attain a higher water speed in order to generate the lift required for takeoff. This increase in water speed means overcoming additional water drag. All of these factors combine to increase takeoff distances and decrease climb performance. In high density altitude conditions, consider not only the length of the water run, but the room required for a safe climbout as well.

The land area around a body of water is invariably somewhat higher than the water surface. Tall trees are common along shorelines, and in many areas, steep or mountainous terrain rises from the water's edge. Be certain the departure path allows sufficient room for safe terrain clearance or for a wide climbing turn back over the water.

There are specific takeoff techniques for different wind and water situations. Large water areas almost always allow a takeoff into the wind, but there are occasionally circumstances where a crosswind or downwind takeoff may be more appropriate. Over the years, techniques have evolved for handling rough water or a glassy smooth surface. Knowing and practicing these techniques not only keep skills polished so they are available when needed, they also increase overall proficiency and add to the enjoyment of seaplane flying.

NORMAL TAKEOFFS
Make normal takeoffs into the wind. Once the wind direction is determined and the takeoff path chosen, configure the seaplane and perform all of the pre-takeoff checks while taxiing to the takeoff position. Verify that the takeoff will not interfere with other traffic, either on the water's surface or in the air.

Hold the elevator control all the way back and apply full power smoothly and quickly, maintaining directional control with the rudder. When the nose reaches its highest point, ease the back pressure to allow the seaplane to come up on the step. Establish the optimum planing attitude and allow the seaplane to accelerate to lift-off speed. In most cases, the seaplane lifts off as it reaches flying speed. Occasionally it may be necessary to gently help the floats unstick by either using some aileron to lift one float out of the water or by adding a small amount of back pressure on the elevator control. Once off the water, the seaplane accelerates more quickly. When a safe airspeed is achieved, establish the pitch attitude for the best rate of climb (V_Y) and complete the climb checklist. Turn as necessary to avoid overflying noise-sensitive areas, and reduce power as appropriate to minimize noise.

CROSSWIND TAKEOFFS
In restricted or limited areas such as canals or narrow rivers, it is not always possible to take off or land directly into the wind. Therefore, acquiring skill in crosswind techniques enhances the safety of seaplane operation. Crosswinds present special difficulties for seaplane pilots. The same force that acts to lift the upwind wing also increases weight on the downwind float, forcing it deeper into the water and increasing drag on that side. Keep in mind that the allowable crosswind component for a floatplane may be significantly less than for the equivalent landplane.

A crosswind has the same effect on a seaplane during takeoff as on a landplane, that is, it tends to push the seaplane sideways across the takeoff path, which imposes side loads on the landing gear. In addition, wind pressure on the vertical tail causes the seaplane to try to weathervane into the wind.

At the beginning of the takeoff roll in a landplane, drift and weathervaning tendencies are resisted by the friction of the tires against the runway, usually assisted by nosewheel steering, or in some cases even differential braking. The objective in a crosswind takeoff is the same in landplanes and seaplanes: to counteract drift and minimize the side loads on the landing gear.

The sideways drifting force, acting through the seaplane's center of gravity, is opposed by the resistance of the water against the side area of the floats. This creates a force that tends to tip the seaplane sideways, pushing the downwind float deeper into the water and lifting the upwind wing. The partly submerged float has even more resistance to sideways motion, and the upwind wing displays more vertical surface area to the wind, intensifying the problem. Without intervention by the pilot, this tipping could continue until the seaplane capsizes.

During a takeoff in stiff crosswinds, weathervaning forces can cause an uncontrolled turn to begin. As the turn develops, the addition of centrifugal force acting outward from the turn aggravates the problem. The keels of the floats resist the sideways force, and the upwind wing tends to lift. If strong enough, the combination of the wind and centrifugal force may tip the seaplane to the point where the downwind float submerges and

subsequently the wingtip may strike the water. This is known as a waterloop, and the dynamics are similar to a groundloop on land. Although some damage occurs when the wingtip hits the ground during a groundloop, the consequences of plunging a wingtip underwater in a seaplane can be disastrous. In a fully developed waterloop, the seaplane may be severely damaged or may capsize. Despite these dire possibilities, crosswind takeoffs can be accomplished safely by exercising good judgment and proper piloting technique.

Since there are no clear reference lines for directional guidance, such as those on airport runways, it can be difficult to quickly detect side drift on water. Waves may make it appear that the water is moving sideways, but remember that although the wind moves the waves, the water remains nearly stationary. The waves are simply an up-and-down motion of the water surface—the water itself is not moving sideways. To maintain a straight path through the water, pick a spot on the shore as an aim point for the takeoff run. On the other hand, some crosswind techniques involve describing a curved path through the water. Experience will help determine which technique is most appropriate for a given situation.

CONTROLLED WEATHERVANING
In light winds, it is easy to counteract the weathervaning tendency during the early part of the takeoff run by creating an allowance for it from the beginning. Prior to adding takeoff power, use the water rudders to set up a heading somewhat downwind of the aim point. The angle will depend on the speed of the wind—the higher

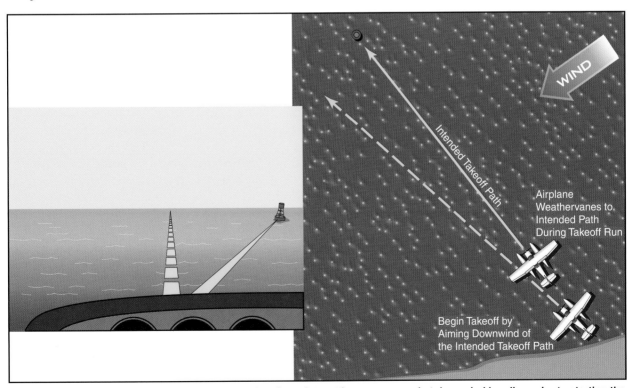

Figure 4-15. Anticipate weathervaning by leading the aim point, setting up a somewhat downwind heading prior to starting the takeoff. Choose an aim point that does not move, such as a buoy or a point on the far shore.

the wind, the greater the lead angle. Create just enough of a lead angle so that when the water rudders are raised and power is applied, the seaplane weathervanes to the desired heading during the time it gains enough speed to make the air rudder and ailerons effective. As the seaplane transitions to the plowing attitude, the weathervaning tendency decreases as the fronts of the floats come out of the water, adding vertical surface area at the front of the seaplane. Use full aileron into the wind as the takeoff run begins, and maintain enough aileron to keep the upwind wing from lifting as airspeed builds. [Figure 4-15 on previous page]

USING WATER RUDDERS
Another technique for maintaining a straight takeoff path involves leaving the water rudders down to assist with steering. Using the water rudders provides added directional control until the aerodynamic controls become effective.

To use this technique, align the seaplane with the aim point on the shore, hold full aileron into the wind, and apply takeoff power. As the seaplane accelerates, use enough aileron pressure to keep the upwind wing down. The downwind float should lift free of the water first. After lift-off, make a coordinated turn to establish the proper crab angle for the climb, and retract the water rudders.

This takeoff technique subjects the water rudders to high dynamic water pressures and could cause damage. Be sure to comply with the advice of the float manufacturer. [Figure 4-16]

DOWNWIND ARC
The other crosswind takeoff technique results in a curved path across the water, starting somewhat into the wind and turning gradually downwind during the takeoff run. This reduces the actual crosswind component at the beginning of the takeoff, when the seaplane is most susceptible to weathervaning. As the aerodynamic controls become more effective, the pilot balances the side loads imposed by the wind with the skidding force of an intentional turn, as always, holding the upwind wing down with the ailerons. [Figure 4-17]

The pilot plans a curved path and follows this arc to produce sufficient centrifugal force so that the seaplane tends to lean outward against the wind force. During the run, the pilot can adjust the rate of turn by varying rudder pressure, thereby increasing or decreasing the centrifugal force to compensate for a changing wind force. In practice, it is quite simple to plan sufficient curvature of the takeoff path to cancel out strong crosswinds, even on very narrow rivers. Note that the

tightest part of the downwind arc is when the seaplane is traveling at slower speeds.

The last portion of a crosswind takeoff is somewhat similar to a landplane. Use ailerons to lift the downwind wing, providing a sideways component of lift to counter the effect of the crosswind. This means that the downwind float lifts off first. Be careful not to drop the upwind wing so far that it touches the water. When using a straight takeoff path, keep the nose on the aim point with opposite rudder and maintain the proper step attitude until the other float lifts off. Unlike a landplane, there is usually no advantage in holding the seaplane on the water past normal lift-off speed, and doing so may expose the floats to unnecessary pounding as they splash through the waves. Once airborne, make a coordinated turn to the crab angle that results in a straight track toward the aim point, and pitch to obtain the desired climb airspeed.

Again, experience plays an important part in successful operation during crosswinds. It is essential that all seaplane pilots have thorough knowledge and skill in these maneuvers.

DOWNWIND TAKEOFFS
Downwind takeoffs in a seaplane present a somewhat different set of concerns. If the winds are light, the water is smooth, and there is plenty of room, a downwind takeoff may be more convenient than a long downwind taxi to a position that would allow a takeoff into the wind. In any airplane, the wing needs to attain a specific airspeed in order to fly, and that indicated airspeed is the same regardless of wind direction.

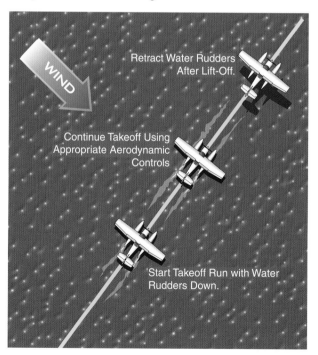

Figure 4-16. Remember to retract the water rudders after takeoff to avoid damage during the next landing.

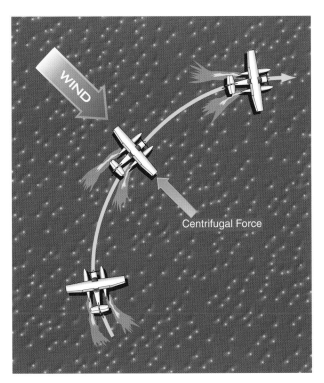

Figure 4-17. The downwind arc balances wind force with centrifugal force.

However, when taking off downwind, obtaining the airspeed means accelerating to a proportionately higher groundspeed. Naturally, the takeoff run is longer because the wings must first be accelerated to the speed of the wind, then accelerated to the correct airspeed to generate the lift required for takeoff. So far, this is identical to what occurs with a landplane during a downwind takeoff. But in addition, a downwind takeoff run in a seaplane is further lengthened by the factor of float drag. The speed of the floats in the water corresponds to the higher groundspeed required in a landplane, but the drag of the floats increases as the square of their speed. This increase in drag is much greater than the increase in rolling resistance of tires and wheel bearings in a landplane. A tailwind may lengthen the seaplane's takeoff distance much more dramatically than the same tailwind in a landplane.

Nevertheless, there are situations in which a downwind takeoff may be more favorable than taking off into the wind. If there is a long lake with mountains at the upwind end and a clear departure path at the other, a downwind takeoff might be warranted. Likewise, noise considerations and thoughtfulness might prompt a downwind takeoff away from a populated shore area if plenty of water area is available. In areas where the current favors a downwind takeoff, the advantage gained from the movement of the water can more than compensate for the wind penalty. Keep in mind that overcoming the current creates far more drag than accelerating a few extra knots downwind with the current. In all cases, safety requires a thorough knowledge of the takeoff performance of the seaplane.

GLASSY WATER TAKEOFFS

Glassy water makes takeoff more difficult in two ways. The smoothness of the surface has the effect of increasing drag, making acceleration and lift-off more difficult. This can feel as if there is suction between the water and the floats. A little surface roughness actually helps break the contact between the floats and the water by introducing turbulence and air bubbles between water and the float bottoms. The intermittent contact between floats and water at the moment of lift-off cuts drag and allows the seaplane to accelerate while still obtaining some hydrodynamic lift, but glassy water maintains a continuous drag force. Once airborne, the lack of visual cues to the seaplane's height above the water can create a potentially dangerous situation unless a positive rate of climb is maintained.

The takeoff technique is identical to a normal takeoff until the seaplane is on the step and nearly at flying speed. At this point, the water drag may prevent the seaplane from accelerating the last few knots to lift-off speed. To reduce float drag and break the grip of the water, the pilot applies enough aileron pressure to lift one float just out of the water and allows the seaplane to continue to accelerate on the step of the other float until lift-off. By allowing the seaplane to turn slightly in the direction the aileron is being held rather than holding opposite rudder to maintain a straight course, considerable aerodynamic drag is eliminated, aiding acceleration and lift-off. When using this technique, be careful not to lift the wing so much that the opposite wing contacts the water. Obviously, this would have serious consequences. Once the seaplane lifts off, establish a positive rate of climb to prevent inadvertently flying back into the water.

Another technique that aids glassy water takeoffs entails roughening the surface a little. By taxiing around in a circle, the wake of the seaplane spreads and reflects from shorelines, creating a slightly rougher surface that can provide some visual depth and help the floats break free during takeoff.

Occasionally a pilot may have difficulty getting the seaplane onto the step during a glassy water takeoff, particularly if the seaplane is loaded to its maximum authorized weight. The floats support additional weight by displacing more water; they sink deeper into the water when at rest. Naturally, this wets more surface area, which equates to increased water drag when the seaplane begins moving, compared to a lightly loaded situation. Under these conditions the seaplane may assume a plowing position when full power is applied, but may not develop sufficient hydrodynamic lift to get on the step due to the additional water drag. The careful seaplane pilot always plans ahead and considers the possibility of aborting the takeoff.

Nonetheless, if these conditions are not too excessive, the takeoff often can be accomplished using the following technique.

After the nose rises to the highest point in the plowing position with full back elevator pressure, decrease back pressure somewhat. The nose will drop if the seaplane has attained enough speed to be on the verge of attaining the step position. After a few seconds, the nose will rise again. At the instant it starts to rise, reinforce the rise by again applying firm back pressure. As soon as the nose reaches its maximum height, repeat the entire routine. After several repetitions, the nose attains greater height and speed increases. If the elevator control is then pushed well forward and held there, the seaplane will slowly flatten out on the step and the controls may then be eased back to the neutral position. Once on the step, the remainder of the takeoff run follows the usual glassy water procedure.

ROUGH WATER TAKEOFFS

The objective in a rough water takeoff is similar to that of a rough or soft field takeoff in a landplane: to transfer the weight of the airplane to the wings as soon as possible, get airborne at a minimum airspeed, accelerate in ground effect to a safe climb speed, and climb out.

In most cases an experienced seaplane pilot can safely take off in rough water, but a beginner should not attempt to take off if the waves are too high. Using the proper procedure during rough water operation lessens the abuse of the floats, as well as the entire seaplane.

During rough water takeoffs, open the throttle to take-off power just as the floats begin rising on a wave. This prevents the float bows from digging into the water and helps keep the spray away from the propeller. Apply a little more back elevator pressure than on a smooth water takeoff. This raises the nose to a higher angle and helps keep the float bows clear of the water.

Once on the step, the seaplane can begin to bounce from one wave crest to the next, raising its nose higher with each bounce, so each successive wave is struck with increasing severity. To correct this situation and to prevent a stall, smooth elevator pressures should be used to set up a fairly constant pitch attitude that allows the seaplane to skim across each successive wave as speed increases. Maintain control pressure to prevent the float bows from being pushed under the water surface, and to keep the seaplane from being thrown into the air at a high pitch angle and low airspeed. Fortunately, a takeoff in rough water is generally accomplished within a short time because if there is sufficient wind to make water rough, the wind is also strong enough to produce aerodynamic lift earlier and enable the seaplane to become airborne quickly.

The relationship of the spacing of the waves to the length of the floats is very important. If the wavelength is less than half the length of the floats, the seaplane is always supported by at least two waves at a time. If the wavelength is longer than the floats, only one wave at a time supports the seaplane. This creates dangerous pitching motions, and takeoff should not be attempted in this situation.

With respect to water roughness, consider the effect of a strong water current flowing against the wind. If the current is moving at 10 knots and the wind is blowing the opposite direction at 15 knots, the relative velocity between the water and the wind is 25 knots, and the waves will be as high as those produced in still water by a wind of 25 knots.

The advisability of canceling a proposed flight because of rough water depends on the size of the seaplane, wing loading, power loading, and, most importantly, the pilot's ability. As a general rule, if the height of the waves from trough to crest is more than half the height of the floats from keel to deck, takeoffs should not be attempted except by expert seaplane pilots. Chapter 8, Emergency Open Sea Operations, contains more information on rough water operations.

CONFINED AREA TAKEOFFS

If operating from a small body of water, an acceptable technique may be to begin the takeoff run while headed downwind, and then turning to complete the takeoff into the wind. This may be done by putting the seaplane on the step while on a downwind heading, then making a step turn into the wind to complete the takeoff. Exercise caution when using this technique since wind and centrifugal force are acting in the same direction and could result in the seaplane tipping over. The water area must be large enough to permit a wide step turn, and winds should be light.

In some cases, the water area may be adequate but surrounding high terrain creates a confined area. The terrain may also block winds, resulting in a glassy water situation as well. Such conditions may lead to a dangerous situation, especially when combined with a high density altitude. Even though landing was not difficult, careful planning is necessary for the takeoff. If the departure path leads over high terrain, consider circling back over the water after takeoff to gain altitude. If air temperatures have increased since landing, make the proper allowance for reduced takeoff performance due to the change in density altitude. Think about spending the night to take advantage of cooler temperatures the next morning. Although the decision may be difficult, consider leaving some cargo or passengers behind if takeoff safety is in question. It is far better to make a second trip to pick them up than to end your takeoff in the trees along the shore.

Chapter 5
Performance

PERFORMANCE CONSIDERATIONS FOR TAKEOFF, CLIMB, CRUISE, AND LANDING

Since many pilots are accustomed to a certain level of performance from a specific make and model of land airplane, the changes in performance when that same airplane is equipped with floats can lead to trouble for a careless or complacent pilot. Floats weigh somewhat more than the wheeled landing gear they replace, but floats are designed to produce aerodynamic lift to offset some of the weight penalty. Generating lift inevitably creates induced drag, which imposes a small reduction in overall performance. By far the greatest impact on performance comes from the parasitic drag of the floats.

TAKEOFF

In a landplane, takeoff distance increases with additional takeoff weight for two reasons: it takes longer for the engine and propeller to accelerate the greater mass to lift-off speed, and the lift-off speed itself is higher because the wings must move faster to produce the extra lift required. For seaplanes, there are two more factors, both due to water drag. As seaplane weight increases, the floats sink deeper into the water, creating more drag during initial acceleration. As with the landplane, the seaplane must also accelerate to a higher airspeed to generate more lift, but the seaplane must overcome significantly more water drag force as speed increases. This extra drag reduces the rate of acceleration and results in a longer takeoff run.

Naturally, the location of the additional weight within the seaplane affects center of gravity (CG) location. Because of the way the floats respond to weight, the CG location affects the seaplane's handling characteristics on the water. If the CG is too far aft, it may be impossible to put the seaplane on the step. If the CG is located to one side of the centerline, one float will be pushed deeper into the water, resulting in more water drag on that side. Be sure to balance the fuel load between left and right wing tanks, and pay attention to how baggage or cargo is secured, so that the weight is distributed somewhat evenly from side to side. [Figure 5-1]

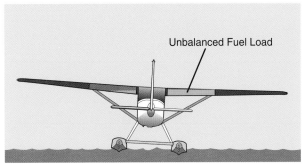

Figure 5-1. The location of the CG can affect seaplane handling.

The importance to weight and balance of pumping out the float compartments should be obvious. Water weighs 8.34 pounds per gallon, or a little over 62 pounds per cubic foot. Performance decreases whenever the wings and engine have to lift and carry useless water in a float compartment. Even a relatively small amount of water in one of the front or rear float compartments could place the airplane well outside of CG limits and seriously affect stability and control. Naturally, water also moves around in response to changes in attitude, and the sloshing of water in the floats can create substantial CG changes as the seaplane is brought onto the step or rotated into a climb attitude.

Some pilots use float compartments near the CG to stow iced fish or game from hunting expeditions. It is imperative to adhere to the manufacturer's weight and balance limitations and to include the weight and moment of float compartment contents in weight and balance calculations.

Density altitude is a very important factor in seaplane takeoff performance. High altitudes, high temperatures, high humidity, and even low barometric pressure can combine to rob the engine and propeller of thrust and the wings of lift. Seaplane pilots are encouraged to occasionally simulate high density altitude by using a reduced power setting for takeoff. This exercise should only be attempted where there is plenty of water area, as the takeoff run will be much longer. An experienced seaplane instructor can assist with choosing an appropriate power setting and demonstrating proper technique.

CLIMB AND CRUISE

When comparing the performance of an airplane with wheels to the same airplane equipped with floats, the drag and weight penalty of the floats usually results in a reduced climb rate for any given weight. Likewise, cruise speeds will usually be a little lower for a particular power setting. This in turn means increased fuel consumption and reduced range. Unless the airplane was originally configured as a seaplane, the performance and flight planning information for a landplane converted to floats will typically be found in the Supplements section rather than the Performance section of the Airplane Flight Manual (AFM) or Pilot's Operating Handbook (POH).

In addition to working within the limits of the seaplane's range, the pilot planning a cross-country flight must also consider the relative scarcity of refueling facilities for seaplanes. Amphibians have access to land airports, of course, but seaplanes without wheels need to find water landing facilities that also sell aviation fuel. While planning the trip, it is wise to call ahead to verify that the facilities have fuel and will be open at the intended arrival times. The Seaplane Pilots Association publishes a Water Landing Directory that is very helpful in planning cross-country flights.

In flight, the seaplane handles very much like the corresponding landplane. On many floatplanes, the floats decrease directional stability to some extent. The floats typically have more vertical surface area ahead of the airplane's CG than behind it. If the floats remain aligned with the airflow, this causes no problems, but if the airplane begins to yaw or skid, this vertical area acts somewhat like a large control surface that tends to increase the yaw, making the skid worse. [Figure 5-2] Additional vertical surface well behind the CG can counteract the yaw force created by the front of the floats, so many floatplanes have an auxiliary fin attached to the bottom of the tail, or small vertical surfaces added to the horizontal stabilizer. [Figure 5-3]

LANDING

Landplane pilots are trained to stay on the lookout for good places to land in an emergency, and to be able to plan a glide to a safe touchdown should the engine(s) fail. An airplane equipped with floats will usually have a steeper power-off glide than the same airplane with wheels. This means a higher rate of descent and a diminished glide range in the event of an engine failure, so the pilot should keep this in mind when spotting potential landing areas during cruising flight.

Seaplanes often permit more options in the event of an unplanned landing, since land can be used as well as water. While a water landing may seem like the only choice for a non-amphibious seaplane, a smooth landing on grass, dirt, or even a hard-surface runway usually causes very little damage to the floats or hull, and may frequently be the safer alternative.

Figure 5-2. The side area of the floats can decrease directional stability.

Figure 5-3. Vertical surfaces added to the tail help restore directional stability.

FLIGHT CHARACTERISTICS OF SEAPLANES WITH HIGH THRUST LINES

Many of the most common flying boat designs have the engine and propeller mounted well above the airframe's CG. This results in some unique handling characteristics. The piloting techniques necessary to fly these airplanes safely are not intuitive and must be learned. Any pilot transitioning to such an airplane is strongly urged to obtain additional training specific to that model of seaplane.

Designing a seaplane with the engine and propeller high above the water offers some important advantages. The propeller is out of the spray during takeoffs and landings, and more of the fuselage volume can be used for passengers and cargo. The pilot usually sits well forward of the wing, and enjoys an excellent view in almost every direction.

Pilots who fly typical light twins are familiar with what happens when one engine is producing power and the other is not. The airplane tends to yaw toward the dead engine. This happens because the thrust line is located some distance from the airplane's CG. In some respects, this situation is similar to the single-engine seaplane with a high thrust line, except that the seaplane flies on one engine all the time. When power is applied, the thrust tends to pitch the nose down, and as power is reduced, the nose tends to rise. [Figure 5-4] This is exactly the opposite of what most pilots are accustomed to. In typical airplanes, including most floatplanes, applying power raises the nose and initiates a climb.

Naturally the magnitude of these pitch forces is proportional to how quickly power is applied or reduced.

The most extreme pitch force logically results from a sudden engine failure, when the full thrust of the engine and its associated downward pitching force are suddenly removed. Forward thrust is replaced by the drag of a windmilling propeller, which adds a new upward pitching force. Since the seaplane is already trimmed with a considerable elevator force to counteract the downward pitch force of the engine, the nose pitches up abruptly. If this scenario occurs just after takeoff, when the engine has been producing maximum power, airspeed is low, and there is little altitude, the pilot must react instantly to overpower the upward pitching forces and push the nose down to avoid a stall.

The reversal of typical pitch forces also comes into play if porpoising should begin during a takeoff. As discussed in Chapter 4, Seaplane Operations - Preflight and Takeoffs, porpoising usually occurs when the planing angle is held too low by the pilot, forcing the front portion of the floats to drag until a wave builds up and travels back along the float. The same thing can happen with the hull of a flying boat, and the nose-down force of a high thrust line can make porpoising more likely. Once porpoising develops, the standard solution is to reduce power and let the airplane settle back into the water. But if power is reduced too quickly in a seaplane with a high-mounted engine, the sudden upward pitching force can combine with the porpoising to throw the seaplane into the air with inadequate airspeed for flight, decreasing thrust, and inadequate altitude for recovery.

Depending on how far the engine is from the airplane's CG, the mass of the engine can have detrimental effects on roll stability. Some seaplanes have the engine mounted within the upper fuselage, while oth-

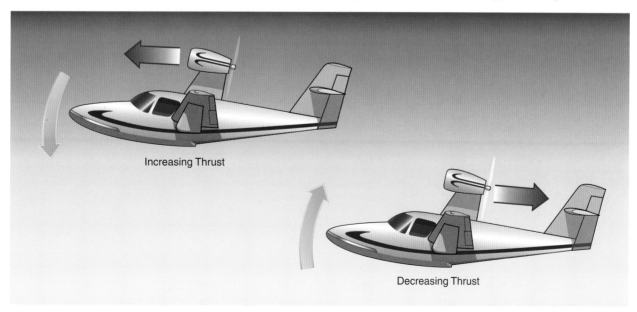

Increasing Thrust

Decreasing Thrust

Figure 5-4. Pitching forces in seaplanes with a high thrust line.

ers have engines mounted on a pylon well above the main fuselage. If it is far from the CG, the engine can act like a weight at the end of a lever, and once started in motion it tends to continue in motion. Imagine balancing a hammer upright with the handle on the palm of the hand. [Figure 5-5]

instructor in order to operate this type of seaplane safely.

MULTIENGINE SEAPLANES

A rating to fly single-engine seaplanes does not entitle a pilot to fly seaplanes with two or more engines. The

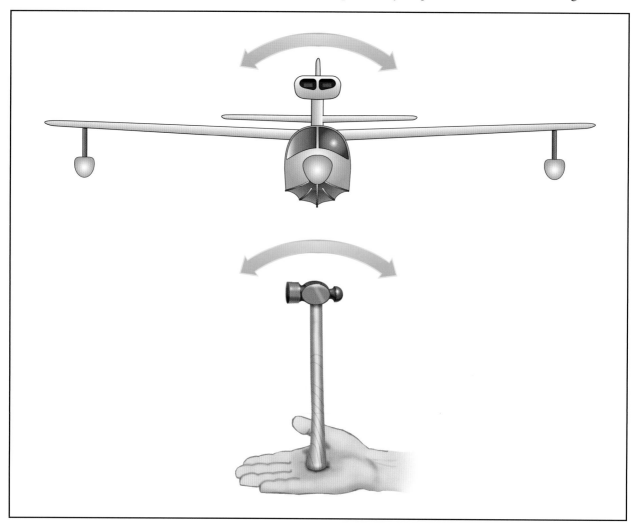

Figure 5-5. Roll instability with a high-mounted engine.

Finally, seaplanes with high-mounted engines may have unusual spin characteristics and recovery techniques. These factors reinforce the point that pilots need to obtain thorough training from a qualified

addition of a multiengine sea rating to a pilot certificate requires considerable additional training. Dealing with engine failures and issues of asymmetrical thrust are important aspects in the operation of multiengine seaplanes.

Chapter 6

Seaplane Operations— Landings

LANDING AREA RECONNAISSANCE AND PLANNING

When a landplane makes an approach at a towered airport, the pilot can expect that the runway surface will be flat and free of obstructions. Wind information and landing direction are provided by the tower. In water operations, the pilot must make a number of judgments about the safety and suitability of the landing area, evaluate the characteristics of the water surface, determine wind direction and speed, and choose a landing direction. It is rare for active airport runways to be used by other vehicles, but common for seaplane pilots to share their landing areas with boats, ships, swimmers, jet-skis, wind-surfers, or barges, as well as other seaplanes.

It is usually a good practice to circle the area of intended landing and examine it thoroughly for obstructions such as pilings or floating debris, and to note the direction of movement of any boats that may be in or moving toward the intended landing site. Even if the boats themselves will remain clear of the landing area, look for wakes that could create hazardous swells if they move into the touchdown zone. This is also the time to look for indications of currents in moving water. Note the position of any buoys marking preferred channels, hidden dangers, or off-limits areas such as no-wake zones or swimming beaches. Just as it is a good idea in a landplane to get a mental picture of the taxiway arrangement at an unfamiliar airport prior to landing, the seaplane pilot should plan a taxi route that will lead safely and efficiently from the intended touchdown area to the dock or mooring spot. This is especially important if there is a significant wind that could make turns difficult while taxiing or necessitate sailing backward or sideways to the dock. If the water is clear, and there is not much wind, it is possible to see areas of waterweeds or obstructions lying below the surface. Noting their position before landing can prevent fouling the water rudders with weeds while taxiing, or puncturing a float on a submerged snag. In confined areas, it is essential to verify before landing that there is sufficient room for a safe takeoff under the conditions that are likely to prevail at the intended departure time. While obstruction heights are regulated in the vicinity of land airports and tall structures are usually well marked, this is not the case with most water landing areas. Be alert for towers, cranes, powerlines, and the masts of ships and boats on the approach path. Finally, plan a safe, conservative path for a go-around should the landing need to be aborted.

Most established seaplane bases have a windsock, but if one is not visible, there are many other cues to gauge the wind direction and speed prior to landing. If there are no strong tides or water currents, boats lying at anchor weathervane and automatically point into the wind. Be aware that some boats also set a stern anchor, and thus do not move with changes in wind direction. There is usually a glassy band of calm water on the upwind shore of a lake. Sea gulls and other waterfowl usually land into the wind and typically head into the wind while swimming on the surface. Smoke, flags, and the set of sails on sailboats also provide the pilot with a fair approximation of the wind direction. If there is an appreciable wind velocity, wind streaks parallel to the wind form on the water. In light winds, they appear as long, narrow, straight streaks of smooth water through the wavelets. In winds of approximately 10 knots or more, foam accents the streaks, forming distinct white lines. Although wind streaks show direction very accurately, the pilot must still determine which end of the wind streak is upwind. For example, an east-west wind streak could mean a wind from the east or the west—it is up to the pilot to determine which. [Figure 6-1]

Figure 6-1. Wind streaks show wind direction accurately, but the pilot must determine which end of the streak is upwind.

If there are whitecaps or foam on top of the waves, the foam appears to move into the wind. This illusion is caused by the motion of the waves, which move more quickly than the foam. As the waves pass under the foam, the foam appears to move in the opposite direction. The shape of shorelines and hills influences wind direction, and may cause significant variations from one area to another. Do not assume that because the wind is from a certain direction on this side of the lake that it is from the same direction on the other side.

Except for glassy water, it is usually best to plan to land on the smoothest water available. When a swell system is superimposed on a second swell system, some of the waves may reinforce each other, resulting in higher waves, while other waves cancel each other out, leaving smoother areas. Often it is possible to avoid the larger waves and land on the smooth areas.

In seaplanes equipped with retractable landing gear (amphibians), it is extremely important to make certain that the wheels are retracted when landing on water. Wherever possible, make a visual check of the wheels themselves, in addition to checking the landing gear position indicators. A wheels-down landing on water is almost certain to capsize the seaplane, and is far more serious than landing the seaplane on land with the wheels up. Many experienced seaplane pilots make a point of saying out loud to themselves before every water landing, "This is a **water** landing, so the wheels should be **up**." Then they confirm that each wheel is up using externally mounted mirrors and other visual indicators. Likewise, they verbally confirm that the wheels are down before every landing on land. The water rudders are also retracted for landings.

When planning the landing approach, be aware that the seaplane has a higher sink rate than its landplane counterpart at the same airspeed and power setting. With some practice, it becomes easy to land accurately on a predetermined spot. Landing near unfamiliar shore-lines increases the possibility of encountering submerged objects and debris.

Besides being safe, it is also very important for seaplane pilots to make a conscious effort to avoid inflicting unnecessary noise on other people in the area. Being considerate of others can often mean the difference between a warm welcome and the banning of future seaplane activity in a particular location. The actions of one pilot can result in the closing of a desirable landing spot to all pilots. People with houses along the shore of a lake usually include the quiet as one of the reasons they chose to live there. Sometimes high terrain around a lake or the local topography of a shoreline can reflect and amplify sound, so that a seaplane sounds louder than it would otherwise. A good practice is to cross populated shorelines no lower than 1,000 feet AGL whenever feasible. To the extent possible consistent with safety, avoid overflying houses during the landing approach. If making a go-around, turn back over the water for the climbout, and reduce power slightly after attaining a safe altitude and airspeed. A reduction of 200 r.p.m. makes a significant difference in the amount of sound that reaches the ground.

LANDING

In water landings, the major objectives are to touch down at the lowest speed possible, in the correct pitch attitude, without side drift, and with full control throughout the approach, landing, and transition to taxiing.

The correct pitch attitude at touchdown in a landplane varies between wide limits. For example, wheel landings in an airplane with conventional-gear, require a nearly flat pitch attitude, with virtually zero angle of attack, while a full-stall landing on a short field might call for a nose-high attitude. The touchdown attitude for a seaplane typically is very close to the attitude for taxiing on the step. The nose may be a few degrees higher. The objective is to touch down on the steps,

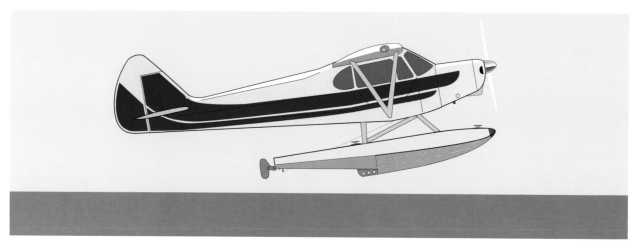

Figure 6-2. The touchdown attitude for most seaplanes is almost the same as for taxiing on the step.

with the sterns of the floats near or touching the water at the same time. [Figure 6-2] If the nose is much higher or lower, the excessive water drag puts unnecessary stress on the floats and struts, and can cause the nose to pitch down, allowing the bows of the floats to dig into the water. Touching down on the step keeps water drag forces to a minimum and allows energy to dissipate more gradually.

NORMAL LANDING

Make normal landings directly into the wind. Seaplanes can be landed either power-off or power-on, but power-on landings are generally preferred because they give the pilot more positive control of the rate of sink and the touchdown spot. To touch down at the slowest possible speed, extend the flaps fully. Use flaps, throttle, and pitch to control the glidepath and establish a stabilized approach at the recommended approach airspeed. The techniques for glidepath control are similar to those used in a landplane.

As the seaplane approaches the water's surface, smoothly raise the nose to the appropriate pitch attitude for touchdown. As the floats contact the water, use gentle back pressure on the elevator control to compensate for any tendency of the nose to drop. When the seaplane is definitely on the water, close the throttle and maintain the touchdown attitude until the seaplane begins to come off the step. Once it begins to settle into the plowing attitude, apply full up elevator to keep the nose as high as possible and minimize spray hitting the propeller.

As the seaplane slows to taxi speed, lower the water rudders to provide better directional control. Raise the flaps and perform the after-landing checklist.

The greater the speed difference between the seaplane and the water, the greater the drag at touchdown, and the greater the tendency for the nose to pitch down. This is why the touchdown is made at the lowest possible speed for the conditions. Many landplane pilots transitioning to seaplanes are surprised at the shortness of the landing run, in terms of both time and distance. It is not uncommon for the landing run from touchdown to idle taxi to take as little as 5 or 6 seconds.

Sometimes the pilot chooses to remain on the step after touchdown. To do so, merely add sufficient power and maintain the planing attitude immediately after touchdown. It is important to add enough power to prevent the seaplane from coming off the step, but not so much that the seaplane is close to flying speed. With too much taxi speed, a wave or swell could throw the seaplane into the air without enough speed to make a controlled landing. In that situation, the seaplane may stall and

contact the water in a nose-down attitude, driving the float bows underwater and capsizing the seaplane. Raising the flaps can help keep the seaplane firmly on the water. To end the step taxi, close the throttle and gradually apply full up elevator as the seaplane slows.

CROSSWIND LANDING

Landing directly into the wind might not be practical due to water traffic in the area, obstructions on or under the water, or a confined landing area, such as a river or canal. In landing a seaplane with any degree of crosswind component, the objectives are the same as when landing a landplane: to minimize sideways drift during touchdown and maintain directional control afterward. Because floats have so much more side area than wheels, even a small amount of drift at touchdown can create large sideways forces. This is important because enough side force can lead to capsizing. Also, the float hardware is primarily designed to take vertical and fore-and-aft loads rather than side loads.

If the seaplane touches down while drifting sideways, the sudden resistance as the floats contact the water creates a skidding force that tends to push the downwind float deeper into the water. The combination of the skidding force, wind, and weathervaning as the seaplane slows down can lead to a loss of directional control and a waterloop. If the downwind float submerges and the wingtip contacts the water when the seaplane is moving at a significant speed, the seaplane could flip over. [Figure 6-3 on next page]

Floatplanes frequently have less crosswind component capability than their landplane counterparts. Directional control can be more difficult on water because the surface is more yielding, there is less surface friction than on land, and seaplanes lack brakes. These factors increase the seaplane's tendency to weathervane into the wind.

One technique sometimes used to compensate for crosswinds during water operations is the same as that used on land; that is, by lowering the upwind wing while holding a straight course with rudder. This creates a slip into the wind to offset the drifting tendency. The apparent movement of the water's surface during the landing approach can be deceiving. Wave motion may make it appear that the water is moving sideways, but although the wind moves the waves, the water itself remains virtually stationary. Waves are simply an up-and-down motion of the water surface—the water itself is not moving sideways. To detect side drift over water and maintain a straight path during landing, pick a spot on the shore or a stationary buoy as an aim point. Lower the upwind wing just enough to stop any drift, and use rudder to maintain a straight

Figure 6-3. Improper technique or excessive crosswind forces can result in an accident.

path. As the seaplane touches down on the upwind float, the water drag will quickly slow the seaplane and the other float will touch down as aerodynamic lift decreases. Close the throttle, and as the seaplane's speed dissipates, increase aileron to hold the upwind wing down. The seaplane is most unstable as it is coming off the step and transitioning through the plowing phase. Be ready for the seaplane to weathervane into the wind as the air rudder becomes less effective. Many pilots make a turn to the downwind side after landing to minimize weathervaning until the seaplane has slowed to taxi speed. Since the seaplane will weathervane sooner or later, this technique reduces the centrifugal force on the seaplane by postponing weathervaning until speed has dissipated. Once the seaplane settles into the displacement attitude, lower the water rudders for better directional control. [Figure 6-4]

Another technique used to compensate for crosswinds (preferred by many seaplane pilots) is the downwind arc method. Seaplanes need not follow a straight path during landing, and by choosing a curved path, the pilot can create a sideward force (centrifugal force) to offset the crosswind force. This is done by steering the seaplane in a downwind arc as shown in figure 6-5. During the approach, the pilot merely plans a curved landing path and follows this path to produce sufficient centrifugal force to counter the wind force. During the landing run, the pilot can adjust the amount of centrifugal force by varying rudder pressure to increase or

decrease the rate of turn. This technique allows the pilot to compensate for a changing wind force during the water run.

Figure 6-5 shows that the tightest curve of the downwind arc is during the time the seaplane is traveling at low speed. Faster speeds reduce the crosswind effect, and at very slow speeds the seaplane can weathervane into the wind without imposing large side loads or stresses. Again, experience plays an important part in successful operation during crosswinds. It is essential that all seaplane pilots have thorough knowledge and skill in these maneuvers.

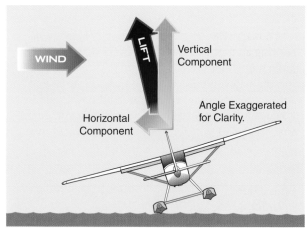

Figure 6-4. Dropping the upwind wing uses a horizontal component of lift to counter the drift of a crosswind.

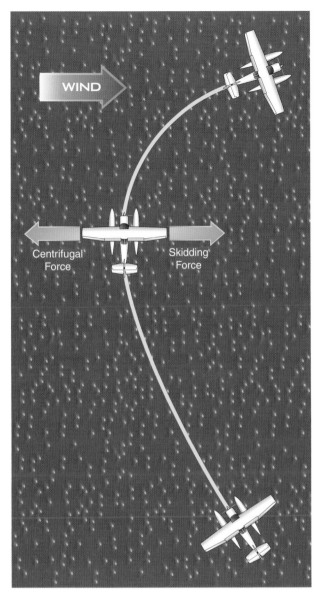

Figure 6-5. A downwind arc is one way to compensate for a crosswind.

DOWNWIND LANDING

Although downwind landings often require significantly more water area, there are occasions when they are more convenient or even safer than landing into the wind. Sometimes landing upwind would mean a long, slow taxi back along the landing path to get to the dock or mooring area. If winds are less than 5 knots and there is ample room, landing downwind could save taxi time. Unless the winds are light, a downwind landing is seldom necessary. Before deciding to land downwind, the pilot needs a thorough knowledge of the landing characteristics of the seaplane as well as the environmental factors in the landing area.

As with a downwind landing in a landplane, the main concern for a seaplane is the additional groundspeed added by the wind to the normal approach speed. The airspeed, of course, is the same whether landing upwind or downwind, but the wind decreases ground-speed in upwind landings and increases groundspeed in downwind landings. While a landplane pilot seldom thinks about the additional force placed on the landing gear by a higher groundspeed at touchdown, it is a serious concern for the seaplane pilot. A small increase in water speed translates into greatly increased water drag as the seaplane touches down, increasing the tendency of the seaplane to nose over. In light winds, this usually presents little problem if the pilot is familiar with how the seaplane handles when touching down at higher speeds, and is anticipating the increased drag forces. In higher winds, the nose-down force may exceed the ability of the pilot or the flight controls to compensate, and the seaplane will flip over at high speed. If the water's surface is rough, the higher touchdown speed also subjects the floats and airframe to additional pounding.

If there is a strong current, the direction of water flow is a major factor in choosing a landing direction. The speed of the current, a confined landing area, or the surface state of the water may influence the choice of landing direction more than the direction of the wind. In calm or light winds, takeoffs usually are made in the same direction as the flow of the current, but landings may be made either with or against the flow of the current, depending on a variety of factors. For example, on a narrow river with a relatively fast current, the speed of the current is often more significant than wind direction, and the need to maintain control of the seaplane at taxi speed after the landing run may present more challenges than the landing itself. It is imperative that even an experienced seaplane pilot obtain detailed information about such operations before attempting them for the first time. Often the best source of information is local pilots with comprehensive knowledge of the techniques that work best in specific locations and conditions.

GLASSY WATER LANDING

Flat, calm, glassy water certainly looks inviting and may give the pilot a false sense of safety. By its nature, glassy water indicates no wind, so there are no concerns about which direction to land, no crosswind to consider, no weathervaning, and obviously no rough water. Unfortunately, both the visual and the physical characteristics of glassy water hold potential hazards for complacent pilots. Consequently, this surface condition is frequently more dangerous than it appears for a landing seaplane.

The visual aspects of glassy water make it difficult to judge the seaplane's height above the water. The lack of surface features can make accurate depth perception very difficult, even for experienced seaplane pilots. Without adequate knowledge of the seaplane's height above the surface, the pilot may flare too high or too low. Either case can lead to an upset. If the seaplane flares too high and stalls, it will pitch down, very likely hitting the water with the bows of the floats and flipping over. If the pilot flares too late or not at all, the seaplane may fly into the water at relatively high speed, landing on the float bows, driving them underwater and flipping the seaplane. [Figure 6-6]

Besides the lack of surface features, the smooth, reflecting surface can lead to confusing illusions as clouds or shore features are reproduced in stunning detail and full color. When the water is crystal clear and glassy, the surface itself is invisible, and pilots may inadvertently judge height by using the bottom of the lake as a reference, rather than the water surface.

The lack of surface texture also presents a physical characteristic that adds slightly to the risk of glassy water landings. A nice smooth touchdown can result in faster deceleration than expected, for the same reason that the floats seem to stick to the surface during glassy water takeoffs: there is less turbulence and fewer air bubbles between the float bottoms and the water, which effectively increases the wetted surface area of the floats and causes higher drag forces. Naturally, this sudden extra drag at touchdown tends to pull the nose down, but if the pilot is expecting it and maintains the planing attitude with appropriate back pressure, the tendency is easily controlled and presents no problem.

There are some simple ways to overcome the visual illusions and increase safety during glassy water landings. Perhaps the simplest is to land near the shoreline, using the features along the shore to gauge altitude. Be certain that the water is sufficiently deep and free of obstructions by performing a careful inspection from a safe altitude. Another technique is to make the final approach over land, crossing the shoreline at the lowest possible safe altitude so that a reliable height reference is maintained to within a few feet of the water surface.

When adequate visual references are not available, make glassy water landings by establishing a stable descent in the landing attitude at a rate that will provide a positive, but not excessive, contact with the water. Recognize the need for this type of landing in ample time to set up the proper final approach. Always perform glassy water landings with power. Perform a normal approach, but prepare as though intending to land at an altitude well above the surface. For example, in a situation where a current altimeter setting is not available and there are few visual cues, this altitude might be 200 feet above the surface. Landing preparation includes completion of the landing checklist and extension of flaps as recommended by the manufacturer. The objective is to have the seaplane ready to contact the water soon after it reaches the target altitude, so at approximately 200 feet above the surface, raise the nose to the attitude normally used for touchdown, and adjust the power to provide a constant descent rate of no more than 150 feet per minute (f.p.m.) at an airspeed approximately 10 knots above stall speed. Maintain this attitude, airspeed, and rate of descent until the seaplane contacts the water. Once the landing attitude and power setting are established, the airspeed and descent rate should remain the same without further adjustment, and the pilot should closely monitor the instruments to maintain this stable glide. Power should only be changed if the airspeed or rate of descent deviate from the desired values. Do not flare, but let the seaplane fly onto the water in the landing attitude. [Figure 6-7]

Upon touchdown, apply gentle back pressure to the elevator control to maintain the same pitch attitude. Close the throttle only after the seaplane is firmly on the water. Three cues provide verification through three different senses—vision, hearing, and body sensation. The pilot sees a slight nose-down pitch at touchdown and perhaps spray thrown to the sides by the floats, hears the sound of the water against the floats, and feels the deceleration force. Accidents have resulted from cutting the power suddenly after the initial touchdown. To the pilot's surprise, a skip had taken place and as the throttle closed, the seaplane was 10 to 15 feet in the air and not on the water, resulting in a stall and substantial damage. Be sure all of the cues

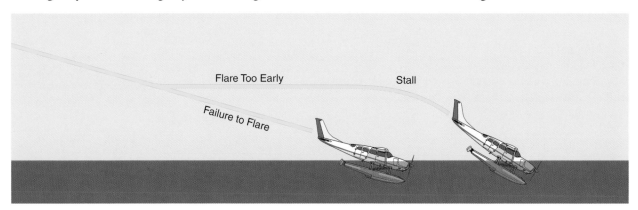

Figure 6-6. The consequences of misjudging altitude over glassy water can be catastrophic.

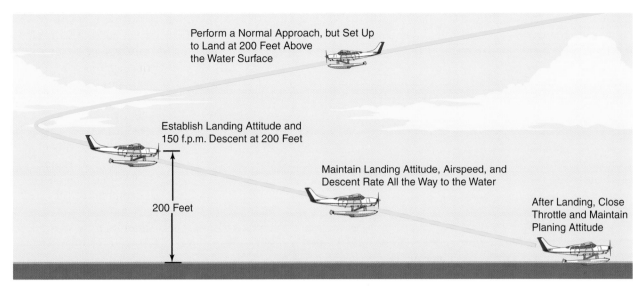

Figure 6-7. Hold the landing attitude, airspeed, and 150 f.p.m. rate of descent all the way to the surface.

indicate that the seaplane is staying on the water before closing the throttle. After the seaplane settles into a displacement taxi, complete the after-landing checklist and lower the water rudders.

An accurately set altimeter may allow the pilot to set up for the touchdown at an altitude somewhat closer to the surface. If the pilot can be certain that the landing configuration and 150 f.p.m. descent will be established well above the water's surface, starting the final glide nearer the surface shortens the descent time and overall landing length.

This technique usually produces a safe, comfortable landing, but the long, shallow glide consumes considerable landing distance. Be certain there is sufficient room for the glide, touchdown, and water run.

ROUGH WATER LANDING

Rough is a very subjective and relative term. Water conditions that cause no difficulty for small boats can be too rough for a seaplane. Likewise, water that poses no challenge to a large seaplane or an experienced pilot may be very dangerous for a smaller seaplane or a less experienced pilot.

Describing a typical or ideal rough water landing procedure is impractical because of the many variables that affect the water's surface. Wind direction and speed must be weighed along with the surface conditions of the water. In most instances, though, make the approach the same as for any other water landing. It may be better, however, to level off just above the water surface and increase the power sufficiently to maintain a rather flat attitude until conditions appear more acceptable, and then reduce the power to touch down. If severe bounces occur, add power and lift off to search for a smoother landing spot.

In general, make the touchdown at a somewhat flatter pitch attitude than usual. This prevents the seaplane from being tossed back into the air at a dangerously low airspeed, and helps the floats to slice through the tops of the waves rather than slamming hard against them. Reduce power as the seaplane settles into the water, and apply back pressure as it comes off the step to keep the float bows from digging into a wave face. If a particularly large wave throws the seaplane into the air before coming off the step, be ready to apply full power to go around.

Avoid downwind landings on rough water or in strong winds. Rough water is usually an indication of strong winds, and vice versa. Although the airspeed for landing is the same, wind velocity added to the seaplane's normal landing speed can result in a much higher groundspeed, imposing excessive stress on the floats, increasing the nose-down tendency at touchdown, and prolonging the water run, since more kinetic energy must be dissipated. As the seaplane slows, the tendency to weathervane may combine with the motion created by the rough surface to create an unstable situation. In strong winds, an upwind landing means a much lower touchdown speed, a shorter water run, and subsequently much less pounding of the floats and airframe.

Likewise, crosswind landings on rough water or in strong winds can leave the seaplane vulnerable to capsizing. The pitching and rolling produced by the water motion increases the likelihood of the wind lifting a wing and flipping the seaplane.

There is additional information on rough water landings in Chapter 8, Emergency Open Sea Operations.

CONFINED AREA LANDING

One of the first concerns when considering a landing in a confined area is whether it is possible to get out again. For most seaplanes, the takeoff run is usually much longer than the landing run. Before landing, the pilot should also consider the wind and surface conditions expected when it is time to leave. If the seaplane lands into a stiff breeze on water with small waves, it might be more difficult to leave the next morning when winds are calm and the water is glassy. Conversely, if the seaplane lands in the morning when the air temperature is low, departure in the hot afternoon might mean a significant loss in takeoff performance due to the density altitude.

It is especially important to carefully inspect the landing area for shallow areas, obstructions, or other hazards. After touchdown is not the time to discover factors that make a confined landing area even smaller or less usable than originally supposed. Evaluation of the landing area should include approach and departure paths. Terrain that rises faster than the seaplane can climb is an obvious consideration, both for the eventual takeoff as well as in case of a go-around during landing. If climbout over the terrain is not easily within the seaplane's capabilities, be certain there is sufficient room to make a gentle turn back over the water for climb.

GO-AROUND

Whenever landing conditions are not satisfactory, execute a go-around. Potential conflicts with other aircraft, surface vessels or swimmers in the landing area, recognition of a hazard on the water, wind shear, wake turbulence, water surface conditions, mechanical failure, or an unstabilized landing approach are a few of the reasons to discontinue a landing attempt. Climb to a safe altitude while executing the go-around checklist, then evaluate the situation, and make another approach under more favorable conditions. Remember that it is often best to make a gentle climbing turn back over the water to gain altitude, rather than climbing out over a shoreline with rising terrain or noise-sensitive areas. The go-around is a normal maneuver that must be practiced and perfected like any other maneuver.

EMERGENCY LANDING

Emergency situations occurring within gliding distance of water usually present no landing difficulty. Although there is some leeway in landing attitude, it is important to select the correct type of landing for the water conditions. If the landing was due to an engine failure, an anchor and paddle are useful after the landing is completed.

Should the emergency occur over land, it is usually possible to land a floatplane with minimal damage in a smooth field. Snow covered ground is ideal if there are no obstructions. The landing should be at a slightly flatter attitude than normal, a bit fast, and directly into the wind. If engine power is available, landing with a small amount of power helps maintain the flatter attitude. Just before skidding to a stop, the tail will begin to rise, but the long front portions of the floats stop the rise and keep the seaplane from flipping over.

A night water landing should generally be considered only in an emergency. They can be extremely dangerous due to the difficulty of seeing objects in the water, judging surface conditions, and avoiding large waves or swell. If it becomes necessary to land at night in a seaplane, seriously consider landing at a lighted airport. An emergency landing can be made on a runway in seaplanes with little or no damage to the floats or hull. Touchdown is made with the keel of the floats or hull as nearly parallel to the surface as possible. After touchdown, apply full back elevator and additional power to lessen the rapid deceleration and nose-over tendency. Do not worry about getting stopped with additional power applied after touchdown. It will stop! The reason for applying power is to provide additional airflow over the elevator to help keep the tail down.

In any emergency landing on water, be as prepared as possible well before the landing. Passengers and crew should put on their flotation gear and adjust it properly. People sitting near doors should hold the liferafts or other emergency equipment in their laps, so no one will need to try to locate or pick it up in the scramble to exit the seaplane. Unlatch all the doors prior to touchdown, so they do not become jammed due to distortion of the airframe. Brief the passengers thoroughly on what to do during and after the landing. These instructions should include how to exit the seaplane even if they cannot see, how to get to the surface, and how to use any rescue aids.

POSTFLIGHT PROCEDURES

After landing, lower the water rudders and complete the after-landing checklist. The flaps are usually raised after landing, both to provide better visibility and to reduce the effects of wind while taxiing. It is a good practice to remain at least 50 feet from any other vessel during the taxi.

After landing, secure the seaplane to allow safe unloading, as well as to keep winds and currents from moving it around. Knowing a few basic terms makes the following discussions easier to understand. Anchoring uses a heavy hook connected to the seaplane by a line or cable. This **anchor** digs into the bottom due to tension on the line, and keeps the seaplane from drifting. **Mooring** means to tie the seaplane to a fixed structure on the surface. The seaplane may be moored to a floating buoy, or to a pier, or to a floating raft. For this discussion, **docking** means securing the seaplane to a permanent structure fixed to the shore. To **beach** a seaplane means to pull it up onto a suitable shore surface, so that its weight is supported by relatively dry ground rather than water. **Ramping** is defined as using a ramp to get the seaplane out of the water and onto the shore.

ANCHORING

Anchoring is the easiest way to secure a seaplane on the water surface. The area selected should be out of the way of moving vessels, and in water deep enough that the seaplane will not be left aground during low tide. The holding characteristics of the bottom are important in selecting an appropriate anchorage. The length of the anchor line should be about seven times the depth of the water. After dropping the anchor with the seaplane headed into the wind, allow the seaplane to drift backward to set the anchor. To be sure the anchor is holding, watch two fixed points somewhere to the side of the seaplane, one farther away than the other, that are aligned with each other, such as a tree on the shore and a mountain in the distance. If they do not remain aligned, it means that the seaplane is drifting and dragging its anchor along the bottom. The nautical term for when two objects appear directly in line, one behind the other, is "in range" and the two objects are called a range.

When choosing a place to anchor, think about what will happen if the wind shifts. Allow enough room so that the seaplane can swing around the anchor without striking nearby obstacles or other anchored vessels. Be certain the water rudders are retracted, as they can interfere with the seaplane's ability to respond to wind shifts.

If anchoring the seaplane overnight or for longer periods of time, use a heavier anchor and be sure to comply with maritime regulations for showing an anchor light or daytime visual signals when required. [Figure 6-8]

When leaving the seaplane anchored for any length of time, it is a good idea to secure the controls with the elevator down and rudder neutral. Since the seaplane can rotate so that it always faces into the wind, this forces the nose down and reduces the angle of attack, keeping lift and wind resistance at a minimum.

MOORING

Mooring a seaplane eliminates the problem of the anchor dragging. A permanent mooring installation consists of a heavy weight on the bottom connected by a chain or cable to a floating buoy with provisions for securing mooring lines. Approach a mooring at a very low speed and straight into the wind. To keep from overrunning the mooring, shut down the engine early and let the seaplane coast to the mooring. If necessary, the engine can be started again for better positioning.

Never straddle a buoy with a twin-float installation. Always approach while keeping the buoy to the outside of the float to avoid damage to the propeller and underside of the fuselage. Initial contact with the buoy is usually made with a boat hook or a person standing on the deck of one float.

While approaching the mooring, have the person on the float secure one end of a short line to the bottom of a float strut, if one is not there already. Then taxi the seaplane right or left of the mooring so that the float on which the person is standing comes directly alongside the buoy. The free end of the line can then be secured to the mooring.

Exercise extreme caution whenever a person is assisting in securing the seaplane. There have been many instances of helpers being struck by the propeller. On

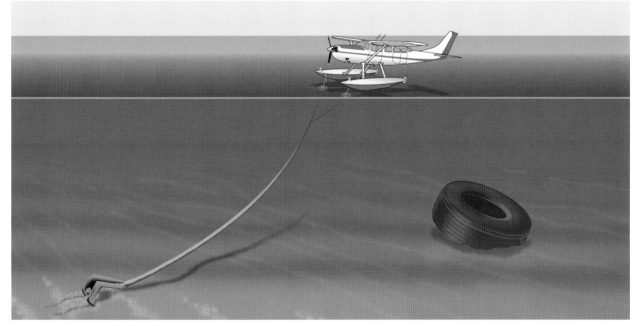

Figure 6-8. Anchoring.

most floatplanes, the floats extend well in front of the propeller arc. Eager to do a good job, an inexperienced helper might forget the spinning propeller while walking forward along the float.

DOCKING

The procedure for docking is essentially the same as for mooring, except that approaching directly into the wind may not be an option. The keys to successful docking are proper planning of the approach to the dock, compensating for the existing environmental conditions, and skill in handling the seaplane in congested areas. Bear in mind that a seaplane is fragile and hitting an obstruction can result in extensive damage.

Plan the approach to the dock so as to keep the wind on the seaplane's nose as much as possible. While still well clear of the dock area, check the responsiveness of the water rudders and be sure the seaplane will be able to maneuver in the existing wind and current. If control seems marginal, turn away and plan an alternative method of reaching the dock. While approaching the dock, the person who will be jumping out to secure the seaplane should take off seatbelts and unlatch the door. When it is clear that the seaplane will just make it to the dock, shut down the engine and let the seaplane coast the remaining distance to encounter the dock as gently as possible. The person securing the seaplane should step out onto the float, pick up the mooring line attached to the rear float strut, and step onto the dock as the seaplane stops. The line should be secured to a mooring cleat on the dock. Use additional mooring lines if the seaplane will be left unattended. Be sure to complete any remaining items on the checklist, and to double-check that the mixture, magnetos, and master switch are in the off positions.

BEACHING

Success in beaching depends primarily on the type and firmness of the shoreline. Inspect the beach carefully before using it. If this is impossible, approach the beach at an oblique angle so the seaplane can be turned out into deeper water if the beach is unsatisfactory. The hardest packed sand is usually near the water's edge and becomes softer where it is dry, further from the water's edge. Rocky shorelines are likely to damage the floats, especially if significant waves are rolling in. Mud bottoms are usually not desirable for beaching.

To protect them from damage, water rudders should be up before entering the shallow water near a beach. Sand is abrasive and erodes any protective coatings on the bottoms of the floats. If possible, beach the seaplane by sailing backward with the water rudders up. The aft bottoms of the floats do not dig into the sand as deeply as the forward bottoms, so backing onto a beach is not as hard on the floats as going in nose-first.

Do not leave the seaplane unattended unless at least a tail line is fastened to some solid object ashore. Moderate action of the water rapidly washes away the sand under the floats and lets the seaplane drift. An incoming tide can float a beached seaplane in just a few minutes. Likewise, a receding tide may leave a seaplane stranded 30 or 40 feet from the water in a few hours. Even small waves may alternately pick up and drop the seaplane, potentially causing serious damage, unless the seaplane is beached well out of their reach. Flying boat pilots should be sure to clear the main gear wells of any sand or debris that may have accumulated before departing.

If the seaplane is beached overnight or higher winds are expected, use portable tiedowns or stakes driven into firm ground and tie it down like a landplane. If severe winds are expected, the compartments of the floats can be filled with water. This holds the seaplane in very high winds, but it is a lot of work to pump out the floats afterward.

RAMPING

For the purpose of this discussion, a ramp is a sloping platform extending well under the surface of the water. If the ramp is wood, the seaplane can be slid up or down it on the keels of the floats, provided the surface of the ramp above the water is wet. Concrete boat ramps are generally not suitable for seaplanes. Water rudders should be down for directional control while approaching the ramp, but raised after the seaplane hits the ramp.

If the wind is blowing directly toward the shore, it is possible to approach the ramp downwind with enough speed to maintain control. Continue this speed until the seaplane actually contacts the ramp and slides up it. Many inexperienced pilots make the mistake of cutting the power before reaching the ramp for fear of hitting it too hard. This is more likely to result in problems, since the seaplane may weathervane and hit the ramp sideways or backward, or at least need to be taxied out for another try. When approaching at the right speed, the bow wave of the float cushions the impact with the ramp, but if the seaplane is too slow or decelerating, the bow wave moves farther back along the float and the impact with the ramp may be harder. Many pilots apply a little power just prior to hitting the ramp, which raises the fronts of the floats and creates more of a cushioning bow wave. Be sure to hold the elevator control all the way back throughout the ramping. [Figure 6-9]

When the seaplane stops moving, shut down the engine and complete the appropriate checklist. Ideally, the seaplane should be far enough up the ramp that waves or swells will not lift the floats and work the seaplane

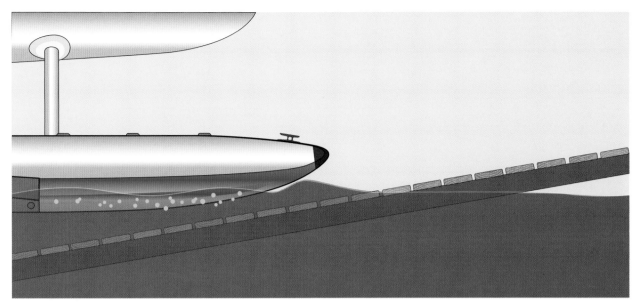

Figure 6-9. The bow wave cushions the contact with the ramp.

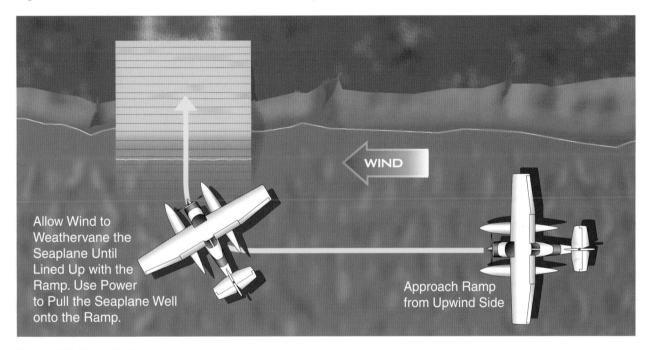

Allow Wind to
Weathervane the
Seaplane Until
Lined Up with the
Ramp. Use Power
to Pull the Seaplane Well
onto the Ramp.

WIND

Approach Ramp
from Upwind Side

Figure 6-10. Crosswind approach to a ramp.

back into the water, but not so far up the ramp that shoving off is difficult. Ramps are usually quite slippery, so pilot and passengers must be very cautious of their footing when walking on the ramp.

The most difficult approach is when the wind is blowing parallel to the shore, and strong enough to make control marginal. If the approach is made into the wind, it may not be possible to turn the seaplane crosswind toward the ramp without excessive speed. In most cases, the best procedure is to taxi directly downwind until near the ramp, then close the throttle at the right point to allow weathervaning to place the seaplane on the ramp in the proper position. Then apply power to pull the seaplane up the ramp and clear of the water. This should not be attempted if the winds are high or the ramp is too slippery, since the seaplane could be blown sideways off the leeward side of the ramp. [Figure 6-10]

Experience and proficiency are necessary for ramping in strong winds. In many instances, the safest procedure is to taxi upwind to the ramp and near enough for a helper to attach a line to the floats. The seaplane may then be left floating, or pushed and pulled into a position where a vehicle can haul it up the ramp.

SALT WATER

Any time the seaplane has been operated in salt water, be sure to flush the entire seaplane with plenty of fresh water to minimize corrosion.

SKIPLANE OPERATIONS

This chapter introduces pilots to the procedures required in the operation of skiplanes. Since most skiplane operations and training are conducted in single-engine airplanes with a conventional gear (tailwheel) configuration, this information is based on operating skiplanes of this type. [Figure 7-1]

Figure 7-1. Skiplane.

A skiplane configuration affects the overall operation and performance of an airplane in several different ways, including ground handling, takeoff, landing, and flight operations. Some manufacturers provide recommended procedures and performance data in the Airplane Flight Manual (AFM) and/or Pilot's Operating Handbook (POH).

Title 14 of the Code of Federal Regulations (14 CFR) part 61 does not require specific pilot training and authorization to operate skiplanes; however, it is important to train with a qualified skiplane flight instructor.

Since most skiplanes operate in a wide variety of conditions, such as landing on frozen or snow-covered lakes and sloping glaciers, with varying qualities of snow, it is important to know how performance is affected. Use the performance data provided by the manufacturer.

CONSTRUCTION AND MAINTENANCE

Modern airplane ski designs are a compromise for the various forms and conditions of snow and ice. For example, a long, wide ski is best for new fallen, powdery, light snow, whereas a sharp, thin blade is best for hard-packed snow or smooth ice. Many ski designs feature a wide, flat ski with aluminum or steel runners on the bottom. Airplane skis may be made from composites, wood, or aluminum, and some have a polyethylene plastic sheathing bonded or riveted to the bottom surfaces. Ski designs fall into two main categories: plain and combination. Plain skis can only be used on snow and ice, while combination skis also allow the wheels to be used to land on runways.

PLAIN SKI TYPES

- **Wheel Replacement**—Wheels are removed and ski boards are substituted. [Figure 7-2]

- **Clamp-On**—Skis that attach to the tires and benefit from the additional shock absorbing qualities of the tires.

- **Roll-On or Full Board**—Similar to the clamp-on type except the tires are bypassed and do not carry side or torque loads. Only the tire cushioning effect is retained with this installation.

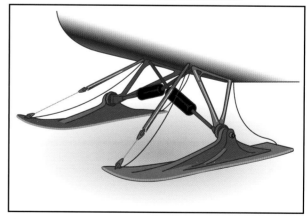

Figure 7-2. Wheel replacement ski.

COMBINATION SKI TYPES

- **Retractable Ski**—Can be extended into place for snow operations or retracted for non-snow operations. This is accomplished by either a hydraulic pump or crank.

- **Penetration Ski**—The wheel extends down partially below the ski, allowing the skiplane to operate from both snow and non-snow surfaces. This type of ski gives poor ground clearance on non-snow surfaces and causes extra drag when on snow. [Figure 7-3]

Figure 7-3. Penetration ski.

The plastic polyethylene sheathing on the bottoms of the skis may be punctured by sharp objects, including ice. It also may shatter from impacts in extremely cold temperatures. Replacing the bonded type sheathing is very difficult in the field. If the sheathing is riveted, machine screws may be used to secure loose sheathing, but the screw holes must provide for expansion and contraction. Follow the manufacturer's recommendations for patching limited sheathing damage. If damage is extensive, the entire ski bottom may need new covering.

Shock cord bungees used in ski rigging deteriorate rapidly when left under tension. When parking the skiplane overnight, detach the bungees at the lower fitting and allow them to hang free. Reattaching the bungees normally requires two or more people.

Hydraulic ski retracting mechanisms usually function well in cold environments, but the small, abrupt change of ski attitude occurring at touchdown imposes a severe load on the external hydraulic lines leading to the skis. These lines are a more prevalent source of trouble than the internal parts.

Use low temperature oils and greases to lubricate friction points. For lubrication requirements, see the AFM/POH or the ski manufacturer's manual.

The condition of the limiting cables and their fastenings is important to safety in flight. Be sure there is no fraying, kinking, rusting, or other defective condition before each flight.

OPERATIONAL CONSIDERATIONS

In the air, skiplane flight characteristics are similar to those of airplanes with standard landing gear, except for a slight reduction in cruising speed and range. Leaving the skis in the extended position in flight produces no adverse effect on trim, but may cause a slight loss of speed. Consult the operator's manual for skiplane performance data, and weight and center of gravity considerations.

The AFM/POH skiplane supplement may provide limitations including limiting airspeeds for operation with skis in flight and for other wheel/ski configurations. These speeds may be different from the wheel-type landing gear configuration, depending on the type of ski and the tension of the springs or bungees holding the fronts of the skis up.

Understand both the limitations and advantages of the ski equipment. Compared to the standard wheel equipped airplane that incorporates individual brakes for steering, skis are clumsy and the airplane is less maneuverable while on the ground. Like a floatplane, a skiplane has a tendency to weathervane with the wind and needs considerable space to maneuver. Maneuvering on the ground and parking require special techniques which are acquired only through practice.

TYPES OF SNOW

- **Powder Snow**—Dry snow in which the water content and ambient temperature are low.

- **Wet Snow**—Contains high moisture and is associated with warmer temperatures near the freezing point.

- **Granular Snow**—Wet snow that has had a temperature drop causing the snow to ball up and/or crust.

TYPES OF ICE

- **Glaze Ice**—Snow that has been packed down and frozen to a solid ice pack, or frozen snow.

- **Glare Ice**—A smooth sheet of ice that is exceedingly slippery with no deformities, cracks, or other irregularities in the surface. This ice lacks any kind of traction, with a coefficient of friction near zero.

- **Clear Ice**—Ice that forms smoothly over a surface and has a transparent appearance.

SURFACE ENVIRONMENTS

- **Glaciers**—Sloping snow or ice packs.

- **Frozen Lakes**—Frozen bodies of water with or without snow cover.

- **Tundra**—A large area of grass clumps supporting snow cover.

PREFLIGHT

Before departing on any trip, it is important to do proper preflight planning. A good preflight should include a review of the proposed route as well as possible alternate routes; terrain; local, en route, and destination weather; fuel requirements; facilities available at the destination; weight and balance; and takeoff and landing distance requirements.

Obtain a complete weather briefing for each leg, and file a flight plan with appropriate remarks. For local flights, always inform someone at home of the area of operation and the expected time of return if a flight plan is not filed.

Include good Aeronautical Decision Making (ADM) procedures, such as running a personal minimums checklist, and think PAVE (Pilot, Aircraft, Environment, and External Pressures) during the preflight phase.

Cold weather is implicit in flying a skiplane, so preflight planning must also include preparations for possible contingencies unique to cold weather operations. This is especially important for flights in bush country, where facilities are scarce and emergency assistance may be limited or nonexistent.

Evaluate all passengers' clothing for suitability in the conditions expected. Consider the passenger when making this evaluation. Children and older people need more protection from the environment than a middle-aged person in good health. Every occupant should be dressed for a long walk, including adequate boots or rubber-bottomed shoes and an arctic parka. Sunglasses are highly recommended, even on cloudy days. Pilots can be blinded by the brightness of the snow, and glare can destroy depth perception.

Survival equipment is required by some states and countries, and many areas require specific items for even the shortest local flights. The requirements usually vary between winter and summer months. Be sure to check the current requirements for the particular jurisdiction. Beyond the minimum requirements, use good judgment to select and carry any other equipment that could help occupants survive an unplanned stay in the specific terrain and environmental conditions along the route of flight. Always consider means of providing warmth, shelter, water, and food; methods of attracting attention and signaling for help in both daylight and darkness; and treatment of injuries. Obtain appropriate survival training and know how to make effective use of the equipment. Whether the cause is a forced landing or an engine that fails to restart after landing at a remote location, the survival gear and clothing should keep the pilot and passengers alive until help arrives.

When planning for an overnight stay away from an airport, or if the skiplane is routinely parked outside, other items may be added to the equipment carried on the skiplane. These might include portable tiedowns, a flashlight, a shovel, and a broom. Wing and fuselage covers can prevent the buildup of frost and snow, simplifying the preflight. In temperatures below 0° F, an engine cover and a catalytic heater may be necessary to preheat the engine compartment. If the pilot carries appropriate hand tools and a bucket, the crankcase oil can be drained from the engine and kept indoors. It may also be helpful to remove the battery and keep it in a warmer location. Many pilots carry burlap sacks, plastic garbage bags, or wooden slats to place under the skis to prevent them from freezing to the surface. Some carry a can of non-stick cooking spray to use on the bottom of the ski to avoid sticking or freezing to the surface. Depending on the needs of the skiplane, it may be necessary to carry extra engine oil, hydraulic fluid, or deicer fluid. Markers such as red rags, colored flags, or glow sticks may come in handy, as well as 50 feet of nylon rope, and an ice pick or ice drill. Select equipment according to the situation, and know how to use it.

If a skiplane has been sitting outside overnight, the most important preflight issues are to ensure that the airframe is free of snow, ice, and frost and that the skis are not frozen to the ground. Often, while sitting on the ground, precipitation may fall and cover the skiplane. Temperatures on the ground may be slightly colder than in the air from which the precipitation falls. When liquid precipitation contacts the colder aircraft structure, it can freeze into a coating of clear ice, which must be removed completely before flight. Wing and tail surfaces must be completely frost free. Any frost, ice, or snow destroys lift and also can cause aileron or elevator flutter. Aerodynamic flutter is extremely dangerous and can cause loss of control or structural failure.

The preflight inspection consists of the standard aircraft inspection and includes additional items associated with the skis. The AFM or POH contains the appropriate supplements and additional inspection criteria. Typical inspection criteria include:

- **Skis**—Examine the skis for damage, delamination, sheathing security, and overall condition.

- **Hardware**—Inspect the condition and security of the clamping bolts, cotter keys, diaper pins, limiting cables, and bungees. Be sure cables and bungees are adjusted properly.

- **Retracting Mechanism**—(if equipped)—Check the hydraulic fluid level and examine the hydraulic lines for leaks. Inspect all cables for fraying and check cable ends for security. Do not cycle the retracting mechanism while on the ground.

- **Ski Freedom**—Be sure the skis are free to move and are not frozen to the surface. If the ambient temperature approaches the melting point, the skis can be freed easily. Gentle swinging of the tail at the rear fuselage, or rocking the airplane at the struts may free the skis. If this does not work, dig the skis out.

- **Tire Pressure**—Check the tire pressure when using skis that depend on the tires for shock absorption, as well as for combination skis. This is especially important if moving a skiplane from a warm hangar to cold temperatures outdoors, as tires typically lose one pound of pressure for every ten-degree drop in ambient temperature.

- **Tailwheel**—Check the tailwheel spring and tail ski for security, cracks, and signs of failure. Without a tail ski, the entire tailwheel and rudder assembly can be easily damaged.

- **Fuel Sump**—During fuel sump checks, sometimes moisture can freeze a drain valve open, allowing fuel to continue to drain. Ice inside the fuel tank could break loose in flight and block fuel lines causing fuel starvation. If the manufacturer recommends the use of anti-icing additives for the fuel system during cold weather operations, follow the ratio and mixing instructions exactly.

- **Survival Equipment**—Check that all required survival equipment is on board and in good condition.

STARTING

Adequately preheat the engine, battery, and the cockpit instruments before startup and departure. Sometimes engine oil may require heating separately. Check the manufacturer's recommendation for starting the engine when ambient temperatures are below freezing.

Batteries require special consideration. In cold climates a strong, fully charged battery is needed. With just a little cold-soaking, the engine may require three times the usual amperage to crank the engine.

Another consideration is the electrolyte freezing point. A fully charged battery can withstand temperatures of -60 to -90° F since the electrolyte's specific gravity is at a

proper level. Conversely, the electrolyte in a weak or discharged battery may freeze at temperatures near 32° F. If a fully charged battery is depleted by an unsuccessful start, it may freeze as it cools to ambient temperature. Later, when the engine is started and the battery is receiving a charge, it could explode.

After start, a proper warmup should be completed prior to a runup and high power settings. Perform the warmup according to the engine manufacturer's recommendations. Some manufacturers recommend a minimum of 1,000 r.p.m. to ensure adequate lubrication.

If the skiplane is parked on heavily crusted snow or glaze ice with the skis frozen to the surface, it may be possible to start the engine and perform the runup in the parking area. Be sure the area behind the skiplane is clear, so as not to cause damage with the propeller wash. If a ski should become unstuck during the runup, reduce power immediately. Then use one of the following procedures to secure the airplane.

Tie down or chock the skiplane prior to engine start, warmup, and runup. Keep all ropes, bags, etc., clear of the propeller. After warmup is complete, and if no assistance is available, shut down the engine to untie and unchock the skiplane, then restart as quickly as possible. If a post, tree, boulder, or other suitable object is available, tie a rope to an accessible structural component in the cockpit, take the end around the anchor object, bring it back to the cockpit, and tie it off with a quick-release knot. When the warmup and runup are complete, release the knot and pull the rope into the cockpit as the skiplane begins to taxi.

If tiedowns or chocks are not available, build small mounds of snow in front of each ski. The mounds must be large enough to prevent the skiplane from taxiing over them during engine start and warmup, but small enough to allow taxiing when power is applied after the warmup is complete. If tiedowns or means to block the skis are not available, the runup can be accomplished while taxiing when clear of obstacles or other hazards. [Figure 7-4]

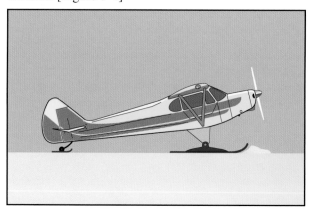

Figure 7-4. Engine warmup.

TAXIING

Taxiing a skiplane on snow and ice presents some unusual challenges. With little or no brakes for stopping or turning, and the ability to skid sideways, a skiplane normally requires more maneuvering room and space to turn than an airplane with wheels.

The tailwheel ski provides marginal directional control on ice and hard packed snow. In such conditions, directional control comes from airflow over the rudder. Adding power and forward elevator control pressure can often help turn the skiplane. The goal is to lighten the tail to help the turn without putting the skiplane on its nose.

Taxiing in strong crosswinds can be difficult. Skiplanes tend to weathervane into the wind. Drifting sideways in the direction of the wind is also commonplace. Taxi in a skid or let the skiplane weathervane partly into the wind during crosswind operations to compensate. [Figure 7-5] A short blast of power may be required to turn the skiplane from upwind to downwind. It is normal to drift sideways in turns. Preplan the taxi track so as to remain clear of drifts, ridges, or other obstructions.

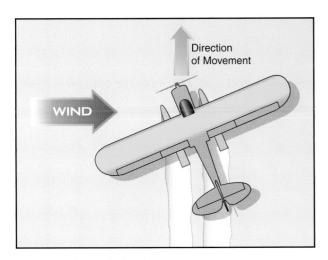

Figure 7-5. Crosswind taxi.

When taxiing in crosswinds on glare ice, get a helper at each wingtip to help with turns and aligning the skiplane for takeoff.

As a general rule, power settings and taxi speeds should be kept as low as possible on ice or crusted snow. On loose or powder snow, add enough power to maintain forward motion and keep the skis on top of the snow. The skiplane may even be step-taxied in a manner similar to a floatplane, staying below takeoff speed. If the skiplane is allowed to sink into soft snow, it may stop moving and become stuck. When the snow is wet and sticky, work the rudder and elevator to get the skiplane moving and maintain forward motion to prevent the skis from sticking again. If the skis are freed during

preflight, but stick again before starting the engine and beginning to taxi, free the skis again and pull the skiplane onto tree branches, leaves, or anything that will prevent the skis from sticking. Burlap bags can be used by tying a line to the bags and pulling them into the cockpit after the skiplane has taxied forward. Keep all ropes, bags, etc., clear of the propeller. Rapid rudder movement will usually break the skis free if they begin to stick during a slow taxi. Use a short blast of power to create more airflow over the tail. A thin coat of engine oil or non-stick cooking spray also prevents sticking if the bottoms of the skis are easily accessed.

At some snow-covered airports, airport managers or fixed base operators spray red or purple dye onto taxi routes and snow banks as visual aids. They may even imbed pine boughs in the snow at regular intervals to help define taxiways and runways or mark hazardous areas. These helpful aids simplify ground operations and improve safety.

TAKEOFFS

Since skiplanes operate from a variety of surfaces, it is important to remember that many takeoff areas can contain unforeseen hazards; therefore, it is important to always plan for the unexpected.

If the condition of the takeoff path is unknown, walk or taxi the full length of the takeoff area and back to check the surface for hazards and help pack the snow. It is better to discover any irregularities before attempting a takeoff than to encounter them at high speeds during takeoff.

Most takeoff distances are greater on snow than for wheel-equipped airplanes on cleared runways and other hard surfaces. On wet or powder snow, two or three times the normal distance may be required. Be sure to remove any frost or crusted snow from the skis before takeoff. Such accumulations increase drag and weight, resulting in a greater takeoff distance.

Select a takeoff direction that provides an adequate distance to lift off and clear any obstructions. Use headwinds or a downhill slope for takeoff when possible to ensure best performance. When turning into the wind, keep moving and turn in a wide arc. Trying to turn too sharply can cause a ski to dig in, resulting in a groundloop or noseover.

Plan and configure for a soft-field takeoff. Soft-field procedures are recommended because the lack of contrast and surface detail or glare off snow or ice may hide possible hazards. Undetected drifts or soft sticky spots can cause sudden deceleration and even a possible noseover.

When lining up to depart, have the skiplane configured properly and keep moving. Do not stop before adding takeoff power because the skiplane may settle into soft snow and limit acceleration. If this happens, it may be necessary to taxi the takeoff path again to pack the snow.

Crosswind takeoffs require the standard procedures and techniques. Be aware that the skiplane may be sliding in a crab during takeoff acceleration. On glaze ice an increase in lateral drift may be seen on takeoff.

OFF-AIRPORT LANDING SITES

Landings on unprepared areas can be accomplished safely if the proper precautions are followed. Evaluating each new landing site thoroughly, obtaining advice from well-qualified pilots already familiar with the area, and staying within the limitations of personal skill and experience can all contribute to safety and reduce risks.

GLACIERS

There are a number of factors that must be considered when operating from glaciers. There can be many hidden hazards.

The first consideration is the condition of the snow and its suitability for landing. To evaluate a new area, fly downhill with the skis on the surface, just touching the snow, as slowly as possible above stall speed. This helps determine the snow condition. If unsure of the quality of the snowpack, look for a gentle slope and land up the slope or hill. This situation will allow the airplane to accelerate easily on a downslope takeoff.

If the slope angle of the landing area is very steep, always evaluate the area for the possibility of an avalanche. Avoid landing near the bottom of a valley because ice falls may exist and provide rough and unusable terrain.

Glaciers are very deceptive. It is advisable to train with an experienced glacier pilot and become comfortable before departing alone. Use extreme caution, as just a few clouds overhead can totally change the picture of the intended landing area.

LAKES AND RIVERS

Snow-covered frozen lakes and rivers can provide a number of obstacles. Wind causes snow to form into ripple-shaped wind drifts. Wind also breaks snow into smaller particles, which bond quickly together to form solid ridges. These ridges can be so rough that they can damage or destroy the landing gear and skiplane. The best plan is to land parallel to ridge rows, even if there is a slight crosswind. Another option is to find a lee area (protected area), where there are no wind drifts and land in this area.

Other problems that may be encountered are beaver dams, houses, or other hidden obstructions that have been covered with snow and have become invisible, especially in flat lighting situations.

A condition known as "overflow" can present problems on landing and takeoff. The overflow is water, in a liquid state, that is cooled below its freezing point. The moment a ski or any other part of the skiplane touches this supercooled water, it freezes solid. As the water freezes, it will provide a rapid deceleration. Overflow may exist on frozen lakes and rivers with or without snow cover. Thin ice also creates a problem because it is not always obvious. It may be thick enough to support a layer of snow or other material, but not a skiplane.

It is easier to see obstacles on lakes and rivers that are frozen without snow cover. Spider holes are ports formed by escaping air from under the ice, forming a weak area or bubble at the surface. These may or may not support the skiplane. Avoid running over spider holes.

Clear ice, under certain conditions, can be extremely slick and will not allow directional control once the aerodynamic controls become ineffective due to the loss of airflow. This becomes critical in crosswind landing conditions.

Avoid landing near the shoreline where rivers or sewer lines empty into lakes. The ice is likely to be very thin in those areas.

TUNDRA

Tundra is probably the least desirable landing surface since most of the above hazards can exist. Tundra is typically composed of small clumps of grass that can support snow and make ridge lines invisible. They also hide obstacles and obscure holes that may be too weak to support skiplanes. Avoid tundra unless the area is well known. [Figure 7-6]

Figure 7-6. Tundra.

LIGHTING

Pilots routinely encounter three general lighting conditions when flying skiplanes. They are flat lighting, whiteout, and nighttime. The implications of nighttime are obvious, and in the interest of safety, night operations from unlighted airstrips are not recommended.

Flat lighting is due to an overcast or broken sky condition with intermittent sunlight. Hills, valleys, and snow mounds take on varying shades of white, and may appear taller, shorter, or wider than they really are. This indirect lighting alters depth perception. The pilot may not realize that depth perception has been compromised, and this can cause serious consequences when operating skiplanes near hilly terrain. When flat lighting is encountered, avoid or discontinue flight operations, especially at an unfamiliar strip.

Whiteout can occur when flying in a valley with both walls obscured by snow or fog. Clear sky conditions can exist, but references cannot be established. Reference to attitude gyro instruments helps when this condition is encountered. Climb out of the valley so additional visual references can be established.

Takeoffs and landings should not be attempted under flat lighting or whiteout conditions.

LANDINGS

Landing a skiplane is easy compared to landing with wheels; however, for off-airport landings, extra precautions are necessary. Be careful in choosing a landing site. Before landing, evaluate the site to be sure a safe departure will be possible.

Upon arriving at a prospective landing site, a pass should be made over the landing area to determine landing direction, and to determine if a safe approach and landing can be completed. A trial landing should be accomplished to determine the best approach, subsequent departure path, and the quality of the surface.

To perform the trial landing, plan and configure for a soft-field landing with a stable approach. Then perform a gentle soft-field touchdown, controlled with power, while remaining near takeoff speed for approximately 600 to 800 feet, and then initiating a go-around.

A trial landing is very helpful in determining the depth and consistency of the snow, evaluating surface conditions, and looking for possible hazards. Be prepared to go around if at any time the landing does not appear normal or if a hazard appears. Do not attempt to land if the ski paths from the trial landing turn black. This indicates "overflow" water beneath the snow wetting the tracks.

When landing on a level surface, and the wind can be determined, make the landing into the wind. If landing on a slope, an uphill landing is recommended. To avoid a hard landing, fly the skiplane all the way to the

surface and add some power just before touchdown. Be sure to turn the skiplane crosswise to the slope before it stops. Otherwise it may slide backward down the slope.

When using combination skis to land on solid ice without the benefit of snow, it is better to land with the wheels extended through the skis to improve the ground handling characteristics. Solid or clear ice surfaces require a much greater landing distance due to the lack of friction. The skiplane also needs more area for turns when taxiing. If the surface has little or no friction, consider the possibility of a groundloop, since the center of gravity is typically behind the main skis and the tail ski may not resist side movement. Keep the skiplane straight during the runout, and be ready to use a burst of power to provide airflow over the rudder to maintain directional control.

Under bright sun conditions and without brush or trees for contrast, glare may restrict vision and make it difficult to identify snowdrifts and hazards. Glare can also impair depth perception, so it is usually best to plan a soft-field landing when landing off airports.

After touchdown on soft snow, use additional power to keep the skiplane moving while taxiing to a suitable parking area and turning the skiplane around. Taxi slowly after landing to allow the skis to cool down prior to stopping. Even though they are moving against cold surfaces, skis warm up a few degrees from the friction and pressure against the surface. Warm skis could thaw the snow beneath them when parked, causing the skis to freeze to the surface when they eventually cool.

PARKING/POSTFLIGHT

Skiplanes do not have any parking brakes and will slide on inclines or sloping surfaces. Park perpendicular to the incline and be prepared to block or chock the skis to prevent movement.

When parked directly on ice or snow, skis may freeze to the surface and become very difficult to free. This happens when there is liquid water under the skis that subsequently freezes. If both the surface and the skis are well below freezing, there will be no problem, but if the skis are warm when the airplane stops, they melt the surface slightly, then the surface refreezes as the heat flows into the ground. Similarly, the weight of the skiplane places pressure on the skis, and pressure generates heat. If the ambient temperature goes up to just below freezing, the heat of pressure can melt the surface under the skis. Then as the temperature drops again, the skis become stuck.

If parking for a considerable amount of time, support the skis above the snow to prevent them from sticking or freezing to the surface. Place tree boughs, wood slats, or other materials under the skis to help prevent them from becoming frozen to the surface. [Figure 7-7] Some pilots apply a coat of non-stick cooking spray or engine oil to the polypropylene ski surface to prevent ice or snow from sticking during the next takeoff.

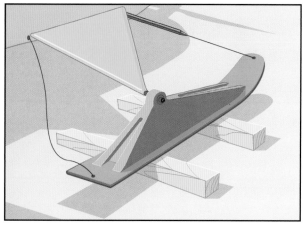

Figure 7-7. Supporting the skis above the surface prevents them from freezing in place.

If the skis are the retractable type and the frozen surface will support the wheels, place the skis in the UP position. Next, dig the snow out from around the skis until ready to depart. This keeps the skis away from the surface. When parking on a hill, pay attention to the position of the fuel selector valve. Typically, the uphill tank should be selected to prevent fuel from transferring to the lower wing and subsequently venting overboard.

EMERGENCY OPERATIONS

When operating a skiplane, carry an adequate survival kit. A good rule of thumb is to carry what is needed to be comfortable. Alaska, Canada, and Sweden provide lists on the internet of the survival equipment required for flights in northern areas. In addition to communicating the current requirements for specific jurisdictions, these lists can help pilots choose additional equipment to meet their needs, beyond the minimum required. Also be sure to check for any restrictions on the carriage of firearms if they are part of your survival kit.

SKI MALFUNCTION

If skis are not rigged properly, or when recommended airspeeds are exceeded, it is possible that a ski will tuck down and give a momentary downward rotation of the nose of the skiplane. This is generally caused by spring or bungee tension not being sufficient to hold ski tips up. The immediate fix is to reduce power and reduce the speed of the skiplane. When the air loads are decreased below the tension of the spring or bungee, the ski will pitch back into place and the control problem will go away. Have a maintenance shop correctly adjust the spring or bungee tension and avoid exceeding the speed limits specified for the skis.

A precautionary landing may be necessary for events such as a broken ski cable or broken hydraulic line. If a ski cable breaks, the front of the ski will tip down. This creates an asymmetrical drag situation, similar to a large speed brake on one side of the skiplane. This condition is controllable; however, it will take skill to maintain control. Not only does the tilted ski create a lot of drag, it also complicates the landing, since the front of the ski will dig in as it contacts the surface, causing abrupt deceleration and severe damage to the landing gear. If efforts to get the ski into a streamlined position fail, a landing should be made as soon as practical.

To attempt to streamline the ski, slow to maneuvering speed or less. It may be possible for a passenger to use a long rod such as a broom handle to push down on the back end of the ski, aligning it with the airflow and making possible a relatively normal landing. If the skis are retractable, try to ensure that they are both in the UP position (for a pavement landing) and land on pavement.

If it is not possible to get the ski to trail correctly, the skiplane must be landed in such a way as to minimize danger to the occupants. This usually means trying to land so that the hanging ski breaks off quickly rather than digging in and possibly destroying the skiplane. Fly to an area where help is available, since damage is virtually inevitable. It is often best to land on a hard surface to increase the chances of the ski breaking away.

With a broken hydraulic line, a condition of one ski up and one ski down may develop. Again, the skiplane is controllable with proper rudder and braking technique.

NIGHT EMERGENCY LANDING

A night landing should never be attempted at an unfamiliar location except in an emergency. To increase the likelihood of a successful landing, perform the checklist appropriate for the emergency, and unlatch the doors prior to landing to prevent jamming due to airframe distortion in the event of a hard landing. If time permits, make distress calls and activate the emergency locator transmitter (ELT).

When selecting a landing area, frozen lakes and rivers are a good choice if the ice is thick enough to support the aircraft. If the ice is thin or the thickness unknown, a landing in an open field would be a better option.

After selecting a landing area, perform a reconnaissance and look for obstructions, field condition, wind direction, and snow conditions if possible. Fly over the landing area in the intended direction of touchdown and drop glow sticks 2 seconds apart along the length of the touchdown zone. Use the glow sticks to aid in depth perception during final approach. Make the touchdown with power, if available, and as slow as possible.

Chapter 8

Emergency Open Sea Operations

OPERATIONS IN OPEN SEAS

Open sea operations are very risky and should be avoided if possible. If an open sea landing cannot be avoided, a thorough reconnaissance and evaluation of the conditions must be performed to ensure safety. The sea usually heaves in a complicated crisscross pattern of swells of various magnitudes, overlaid by whatever chop the wind is producing. A relatively smooth spot may be found where the cross swells are less turbulent. Both a high and a low reconnaissance are necessary for accurate evaluation of the swell systems, winds, and surface conditions.

DEFINITIONS

When performing open sea operations, it is important to know and understand some basic ocean terms. A thorough knowledge of these definitions allows the pilot to receive and understand sea condition reports from other aircraft, surface vessels, and weather services.

Fetch—An area where wind is generating waves on the water surface. Also the distance the waves have been driven by the wind blowing in a constant direction without obstruction.

Sea—Waves generated by the existing winds in the area. These wind waves are typically a chaotic mix of heights, periods, and wavelengths. Sometimes the term refers to the condition of the surface resulting from both wind waves and swells.

Swell—Waves that persist outside the fetch or in the absence of the force that generated them. The waves have a uniform and orderly appearance characterized by smooth, regularly spaced wave crests.

Primary Swell—The swell system having the greatest height from trough to crest.

Secondary Swells—Swell systems of less height than the primary swell.

Swell Direction—The direction from which a swell is moving. This direction is not necessarily the result of the wind present at the scene. The swell encountered may be moving into or across the local wind. A swell tends to maintain its original direction for as long as it continues in deep water, regardless of changes in wind direction.

Swell Face—The side of the swell toward the observer. The back is the side away from the observer.

Swell Length—The horizontal distance between successive crests.

Swell Period—The time interval between the passage of two successive crests at the same spot in the water, measured in seconds.

Swell Velocity—The velocity with which the swell advances in relation to a fixed reference point, measured in knots. (There is little movement of water in the horizontal direction. Each water particle transmits energy to its neighbor, resulting primarily in a vertical motion, similar to the motion observed when shaking out a carpet.)

Chop—A roughened condition of the water surface caused by local winds. It is characterized by its irregularity, short distance between crests, and whitecaps.

Downswell—Motion in the same direction the swell is moving.

Upswell—Motion opposite the direction the swell is moving. If the swell is moving from north to south, a seaplane going from south to north is moving upswell.

SEA STATE EVALUATION

Wind is the primary cause of ocean waves and there is a direct relationship between speed of the wind and the state of the sea in the immediate vicinity. Windspeed forecasts can help the pilot anticipate sea conditions. Conversely, the condition of the sea can be useful in determining the speed of the wind. Figure 8-1 on the next page illustrates the **Beaufort wind scale** with the corresponding **sea state condition number**.

While the height of the waves is important, it is often less of a consideration than the wavelength, or the distance between swells. Closely spaced swells can be very violent, and can destroy a seaplane even though the wave height is relatively small. On the other hand, the same seaplane might be able to handle much higher waves if the swells are several thousand feet apart. The relationship between the swell length and the height of

BEAUFORT WIND SCALE WITH CORRESPONDING SEA STATE CODES

Beaufort Number	Wind Velocity (Knots)	Wind Description	Sea State Description	Sea State Term and Height of Waves (Feet)	Sea State Condition Number
0	Less than 1	Calm	Sea surface smooth and mirror-like	Calm, glassy 0	0
1	1-3	Light Air	Scaly ripples, no foam crests		
2	4-6	Light Breeze	Small wavelets, crests glassy, no breaking	Calm, rippled 0 – 0.3	1
3	7-10	Gentle Breeze	Large wavelets, crests begin to break, scattered whitecaps	Smooth, wavelets 0.3-1	2
4	11-16	Moderate Breeze	Small waves, becoming longer, numerous whitecaps	Slight 1-4	3
5	17-21	Fresh Breeze	Moderate waves, taking longer form, many whitecaps, some spray	Moderate 4-8	4
6	22-27	Strong Breeze	Larger waves, whitecaps common, more spray	Rough 8-13	5
7	28-33	Near Gale	Sea heaps up, white foam streaks off breakers	Very rough 13-20	6
8	34-40	Gale	Moderately high, waves of greater length, edges of crests begin to break into spindrift, foam blown in streaks		
9	41-47	Strong Gale	High waves, sea begins to roll, dense streaks of foam, spray may reduce visibility		
10	48-55	Storm	Very high waves, with overhanging crests, sea white with densely blown foam, heavy rolling, lowered visibility	High 20-30	7
11	56-63	Violent Storm	Exceptionally high waves, foam patches cover sea, visibility more reduced	Very high 30-45	8
12	64 and over	Hurricane	Air filled with foam, sea completely white with driving spray, visibility greatly reduced	Phenomenal 45 and over	9

Figure 8-1. Beaufort wind scale.

the waves is the **height-to-length ratio** [Figure 8-2]. This ratio is an indication of the amount of motion a seaplane experiences on the water and the threat to capsizing. For example, a body of water with 20-foot waves and a swell length of 400 feet has a height-to-length ratio of 1:20, which may not put the seaplane at risk of capsizing, depending on the crosswinds.

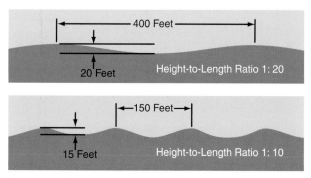

Figure 8-2. Height-to-length ratio.

However, 15-foot waves with a length of 150 feet produce a height-to-length ratio of 1:10, which greatly increases the risk of capsizing, especially if the wave is breaking abeam of the seaplane. As the swell length decreases, swell height becomes increasingly critical to capsizing. Thus, when a high swell height-to-length ratio exists, a crosswind takeoff or landing should not be attempted. Downwind takeoff and landing may be made downswell in light and moderate wind; however, a downwind landing should never be attempted when wind velocities are high regardless of swell direction.

When two swell systems are in phase, the swells act together and result in higher swells. However, when two swell systems are in opposition, the swells tend to cancel each other or "fill in the troughs." This provides a relatively flat area that appears as a lesser concentration of whitecaps and shadows. This flat area is a good touchdown spot for landing. [Figure 8-3]

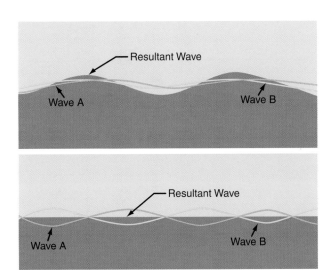
Figure 8-3. Wave interference.

SWELL SYSTEM EVALUATION

The purpose of the swell system evaluation is to determine the surface conditions and the best heading and technique for landing. Perform a high reconnaissance, a low reconnaissance, and then a final determination of landing heading and touchdown area.

HIGH RECONNAISSANCE

During the high reconnaissance, determine the swell period, swell velocity, and swell length. Perform the high reconnaissance at an altitude of 1,500 to 2,000 feet. Fly straight and level while observing the swell systems. Perform the observation through a complete 360° pattern, rolling out approximately every 45°.

Fly parallel to each swell system and note the heading, the direction of movement of the swell, and the direction of the wind.

To determine the time and distance between crests, and their velocity, follow these directions:

1. Drop smoke or a float light and observe the wind condition.

2. Time and count the passage of the smoke or float light over successive crests. The number of waves is the number of crests counted minus one. (A complete wave runs from crest to crest. Since the timing starts with a crest and ends with a crest, there is one less wave than crests.) Time and count each swell system.

3. Obtain the swell period by dividing the time in seconds by the number of waves. For example, 5 waves in 30 seconds equates to a swell period of 6 seconds.

4. Determine the swell velocity in knots by multiplying the swell period by 3. In this example, 6 seconds multiplied by 3 equals 18 knots.

5. To determine the swell length or distance between crests in feet, multiply the square of the swell period by 5. For example, using a 6-second swell period, 6^2 multiplied by 5 equals 180 feet. [Figure 8-4]

Swell Period	$\dfrac{\text{Time in Seconds}}{\text{Number of Waves Counted}}$
Swell Velocity	Swell Period x 3 knots
Swell Length	Swell Period^2 x 5 Feet

Figure 8-4. Rules of thumb to determine swell period, velocity, and length.

LOW RECONNAISSANCE

Perform the low reconnaissance at 500 feet to confirm the findings of the high reconnaissance and obtain a more accurate estimate of wind direction and velocity.

If the direction of the swell does not agree with the direction noted at 2,000 feet, then there are two swell systems from different directions. The secondary swell system is often moving in the same direction as the wind and may be superimposed on the first swell system. This condition may be indicated by the presence of periodic groups of larger-than-average swells.

The wind direction and speed can be determined by dropping smoke or observing foam patches, whitecaps, and wind streaks. Whitecaps fall forward with the wind but are overrun by the waves. Thus, the foam patches appear to slide backward into the direction from which the wind is blowing. To estimate wind velocity from sea surface indications, see figure 8-1.

SELECT LANDING HEADING

When selecting a landing heading, chart all observed variables and determine the headings that will prove the safest while taking advantage of winds, if possible. Descend to 100 feet and make a final evaluation by flying the various headings and note on which heading the sea appears most favorable. Use the heading that looks smoothest and corresponds with one of the possible headings selected by other criteria.

Consider the position of the sun. A glare on the water during final approach might make that heading an unsafe option.

Use caution in making a decision based on the appearance of the sea. Often a flightpath directly downswell appears to be the smoothest, but a landing on this heading could be disastrous.

SELECT TOUCHDOWN AREA

On final approach, select the touchdown area by searching for a null or smooth area in the swell system, avoiding rough areas if possible. When doing so, consider the conditions discussed in the following sections.

LANDING PARALLEL TO THE SWELL

When landing on a swell system with large, widely spaced crests more than four times the length of the floats, the best landing heading parallels the crests and has the most favorable headwind component. In this situation, it makes little difference whether touchdown is on top of the crest or in the trough.

LANDING PERPENDICULAR TO THE SWELL

If crosswind limits would be exceeded by landing parallel to the swell, landing perpendicular to the swell might be the only option. Landing in closely spaced swells less than four times the length of the floats should be considered an emergency procedure only, since damage or loss of the seaplane can be expected. If the distance between crests is less than half the length of the floats, the touchdown may be smooth, since the floats will always be supported by at least two waves, but expect severe motion and forces as the seaplane slows.

A downswell landing on the back of the swell is preferred. However, strong winds may dictate landing into the swell. To compare landing downswell with landing into the swell, consider the following example. Assuming a 10-second swell period, the length of the swell is 500 feet, and it has a velocity of 30 knots or 50 feet per second. Assume the seaplane takes 890 feet and 5 seconds for its runout.

Downswell Landing—The swell is moving with the seaplane during the landing runout, thereby increasing the effective swell length by about 250 feet and resulting in an effective swell length of 750 feet. If the seaplane touches down just beyond the crest, it finishes its runout about 140 feet beyond the next crest. [Figure 8-5]

Landing into the Swell—During the 5 seconds of runout, the oncoming swell moves toward the seaplane a distance of about 250 feet, thereby shortening the effective swell length to about 250 feet. Since the seaplane takes 890 feet to come to rest, it would meet the oncoming swell less than halfway through its runout and it would probably be thrown into the air, out of control. Avoid this landing heading if at all possible. [Figure 8-6]

If low ceilings prevent complete sea evaluation from the altitudes prescribed above, any open sea landing should be considered a calculated risk, as a dangerous but unobserved swell system may be present in the proposed landing area. Complete the descent and before-landing checklists prior to descending below 1,000 feet if the ceiling is low.

LANDING WITH MORE THAN ONE SWELL SYSTEM

Open water often has two or more swell systems running in different directions, which can present a confusing appearance to the pilot. When the secondary swell system is from the same direction as the wind, the preferred direction of landing is parallel to the primary swell with the secondary swell at some angle. When landing parallel to the primary swell, the two choices of heading are either upwind and into the secondary swell, or downwind and downswell. The heading with the greatest headwind is preferred; however, if a pronounced secondary swell system is present, it may be desirable to land downswell to the secondary swell system and accept some tailwind component. The risks associated with landing downwind versus downswell must be carefully considered. The choice of heading depends on the velocity of the wind versus the velocity and the height of the secondary swell. [Figure 8-7]

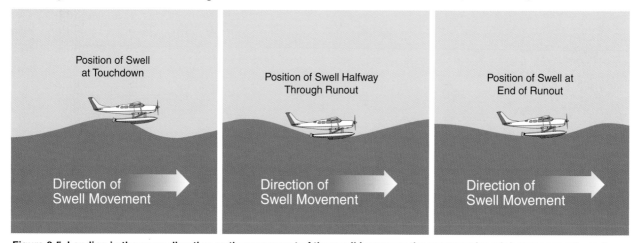

Figure 8-5. Landing in the same direction as the movement of the swell increases the apparent length between swell crests.

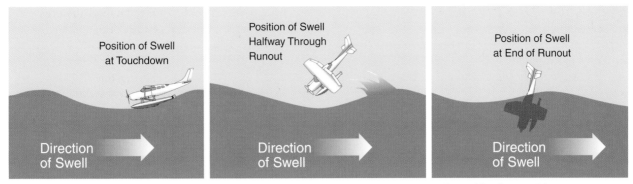

Figure 8-6. Landing against the swell shortens the apparent distance between crests, and could lead to trouble.

Due to the rough sea state, landings should not be attempted in winds greater than 25 knots except in extreme emergencies. Crosswind limitations for each type of seaplane must be the governing factor in crosswind landings.

EFFECT OF CHOP

Chop consists of small waves caused by local winds in excess of 14 knots. These small waves ride on top of the swell system and, if severe, may hide the underlying swell system. Alone, light and moderate chop are not considered dangerous for landings.

NIGHT OPERATIONS

Night landings in seaplanes on open water are extremely dangerous with a high possibility of damage or loss of the seaplane. A night landing should only be performed in an extreme emergency when no other options are available. A night landing on a lighted runway exposes the seaplane to much less risk.

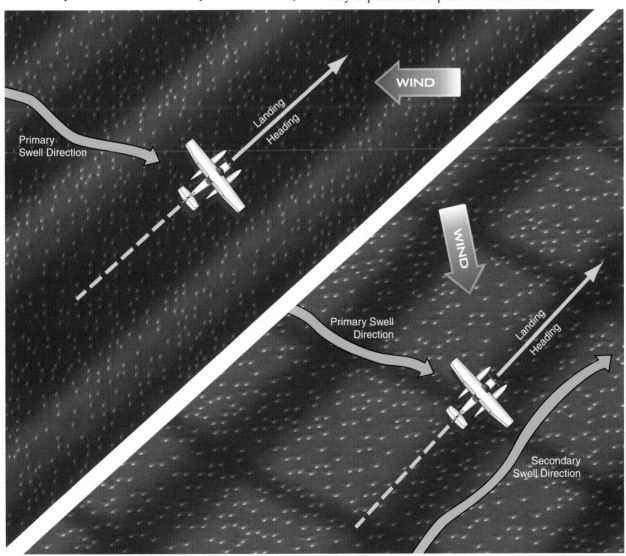

Figure 8-7. Landing heading in single and multiple swell systems.

If operating at night, equip the seaplane with parachute flares, smoke floats, glow sticks, or other markers.

SEA EVALUATION AT NIGHT

Before attempting a night landing, perform a sea state evaluation as described in previous sections. If an emergency occurs shortly after nightfall, a landing heading can be determined by estimating the current conditions from those conditions prevalent before nightfall. If the pilot has no information to form an estimate of the conditions, the information must be obtained from other sources or determined by the pilot from a sea state evaluation by flare illumination or moonlight. If near a ship, sea weather conditions and a recommended landing heading may be obtained from the ship. However, a landing heading based on such information is subject to error and should only be used as a last resort. A pilot evaluation is preferred and can be accomplished by performing the teardrop pattern night sea evaluation as follows:

1. Set a parachute flare and adjust the altitude so that the flare ignites at 1,700 feet. Altitude should be as close to 2,000 feet as possible.

2. After the drop, adjust altitude to 2,000 feet and maintain the heading for 45 seconds.

3. Turn back 220°, left or right, until the flare is almost dead ahead. The sea becomes visible after the first 70° of the turn is completed, allowing approximately 90 seconds for sea evaluation. Use standard rate turn (3° per second).

4. Immediately after passing the flare, if it is still burning, the pilot may circle to make additional evaluation during remaining burning time.

If both pilot and copilot are present, the pilot should fly the seaplane and the copilot should concentrate on the sea evaluation. If only two flares are available and sea conditions are known or believed to be moderate, it may be advisable to dispense with the sea evaluation and use both flares for landing.

NIGHT EMERGENCY LANDING

A night landing should be performed only after exhausting all other options. Be sure all occupants are wearing life vests and secure loose items prior to touchdown. Remove liferafts and survival equipment from their storage containers and give them to those occupants closest to the exits. Prior to the landing pattern, unlatch the doors to prevent jamming that may be caused by airframe distortion from a hard landing. If time permits, make distress calls and activate the emergency locator transmitter.

LANDING BY PARACHUTE FLARE

When a landing heading has been determined and all emergency and cockpit procedures have been

accomplished, the landing approach with the use of parachute flares is made as follows:

1. Establish a heading 140° off the selected landing heading.

2. Lower the flaps and establish the desired landing pattern approach speed.

3. As close to 2,000 feet above the surface as possible, set the parachute flare and adjust the altitude so the flare ignites at 1,700 feet.

4. Release the flare and begin a descent of 900 f.p.m. while maintaining heading for 45 seconds. If the starting altitude is other than 2,000 feet, determine the rate of descent by subtracting 200 feet and dividing by two. (For example, 1800 feet minus 200 is 1600, divided by 2 equals an 800 f.p.m. rate of descent).

5. After 45 seconds, make a standard rate turn of 3° per second toward the landing heading in line with the flare. This turn is 220° and takes approximately 73 seconds.

6. Roll out on the landing heading in line with the flare at an altitude of 200 feet. During the last two-thirds of the turn, the water is clearly visible and the seaplane can be controlled by visual reference.

7. Land straight ahead using the light of the flare. Do not overshoot. Overshooting the flare results in a shadow in front of the aircraft making depth perception very difficult. The best touchdown point is several hundred yards short of the flare.

A rapid descent in the early stages of the approach allows a slow rate of descent when near the water. This should prevent flying into the water at a high rate of descent due to faulty depth perception or altimeter setting. [Figure 8-8]

LANDING BY MARKERS

If parachute flares are not available, use a series of lighted markers to establish visual cues for landing. When a landing heading has been determined and all emergency and cockpit procedures are completed, use drift signals or smoke floats and perform the landing approach as follows:

1. Establish a heading on the reciprocal of the landing heading.

2. Drop up to 20 markers at 2 second intervals.

3. Perform a right 90° turn followed immediately by a 270° left turn while descending to 200 feet.

4. Slightly overshoot the turn to the final approach heading to establish a path parallel and slightly to the right of the markers.

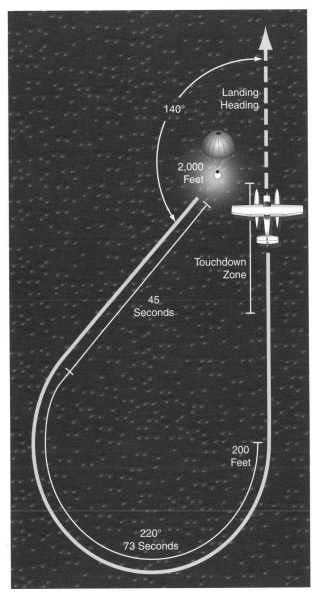

Figure 8-8. Landing by parachute flare.

5. Establish a powered approach with a 200 f.p.m. rate of descent and airspeed 10 percent to 20 percent above stall speed with flaps down, as if for a glassy water landing.

6. Maintain the landing attitude until water contact, and reduce power to idle after touchdown.

Do not use landing lights during the approach unless considerable whitecaps are present. The landing lights may cause a false depth perception. [Figure 8-9]

EMERGENCY LANDING UNDER INSTRUMENT CONDITIONS

When surface visibilities are near zero, the pilot has no alternative but to fly the seaplane onto the water by instruments. A landing heading can be estimated from forecasts prior to departure, broadcast sea conditions, or reports from ships in the area. Obtain the latest local altimeter setting to minimize the possibility of altitude errors during the approach.

Due to the high possibility of damage or capsizing upon landing, be sure all occupants have life vests on and secure all loose items prior to touchdown. Remove liferafts and survival equipment from their storage containers and give them to those occupants closest to the exits. Prior to the landing pattern, unlatch doors to prevent jamming caused by airframe distortion from a hard landing. If time permits, transmit a distress call and activate the emergency locator transmitter.

After choosing a landing heading, establish a final approach with power and set up for a glassy water landing. Establish a rate of descent of 200 f.p.m. and maintain airspeed 10 to 20 percent above stall speed with flaps down. Establish the landing attitude by referring to the instruments. Maintain this approach until the seaplane makes contact with the water, or until visual contact is established.

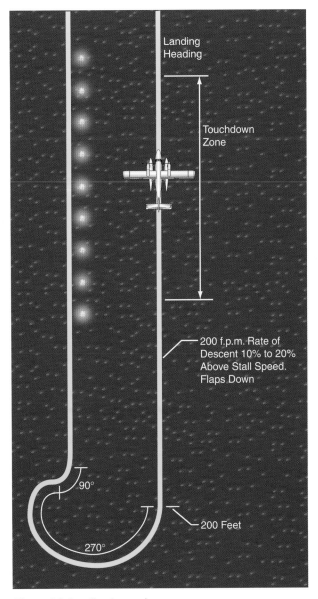

Figure 8-9. Landing by markers.

ESCAPING A SUBMERGED SEAPLANE

If a seaplane capsizes, it is absolutely essential that both pilot and passengers understand how to exit the seaplane and find their way safely to the surface. Pilots should become thoroughly familiar with possible escape scenarios and practice to the extent possible so that they will be able to react instantly in an emergency. Passengers can not be expected to have any prior training in water survival, and an actual emergency is not a good time to try to instruct them. Therefore, a complete briefing before takeoff is very important. At a minimum, the portions of the passenger briefing that deal with escaping from the seaplane in an emergency should cover orientation, water pressure issues, the use of flotation equipment, and both normal and unusual methods of leaving the seaplane.

ORIENTATION

Many of those who have survived seaplane accidents emphasize how disorienting this situation can be. Unlike the clear water of a swimming pool, the water around a seaplane after an accident is usually murky and dark, and may be nearly opaque with suspended silt. In most cases the seaplane is in an unusual attitude, making it difficult for passengers to locate doors or emergency exits. In a number of cases, passengers have drowned while pilots have survived simply because of the pilots' greater familiarity with the inside of the seaplane. Use the preflight briefing to address disorientation by helping passengers orient themselves regardless of the seaplane's attitude. Help the passengers establish a definite frame of reference inside the seaplane, and remind them that even if the cabin is inverted, the doors and exits remain in the same positions relative to their seats. Also, brief passengers on how to find their way to the surface after getting clear of the seaplane. Bubbles always rise toward the surface, so advise passengers to follow the bubbles to get to the surface.

WATER PRESSURE

The pressure of water against the outside of the doors and windows may make them difficult or impossible to open. Passengers must understand that doors and windows that are already underwater may be much easier to open, and that it may be necessary to equalize the pressure on both sides of a door or window before it will open. This means allowing the water level to rise or flooding the cabin adjacent to the door, which can be very counter-intuitive when trapped underwater.

FLOTATION EQUIPMENT

Personal flotation devices (PFDs) are highly recommended for pilots and all passengers on seaplanes.

Since the probability of a passenger finding, unwrapping, and putting on a PFD properly during an actual capsizing is rather low, some operators encourage passengers to wear them during the starting, taxiing, takeoff, landing, and docking phases of flight.

Not all PFDs are appropriate for use in aircraft. Those that do not have to be inflated, and that are bulky and buoyant all the time, can be more of a liability in an emergency, and actually decrease the wearer's chances of survival. Many of the rigid PFDs used for water recreation are not suitable for use in a seaplane. In general, PFDs for aircraft should be inflatable so that they do not keep the user from fitting through small openings or create buoyancy that could prevent the wearer from swimming downward to an exit that is underwater. Obviously, once the wearer is clear of the seaplane, the PFD can be inflated to provide ample support on the water.

The pretakeoff briefing should include instructions and a demonstration of how to put on and adjust the PFD, as well as how to inflate it. It is extremely important to warn passengers never to inflate the PFD inside the seaplane. Doing so could impede their ability to exit, prevent them from swimming down to a submerged exit, risk damage to the PFD that would make it useless, and possibly block the exit of others from the seaplane.

NORMAL AND UNUSUAL EXITS

The briefing should include specifics of operating the cabin doors and emergency exits, keeping in mind that this may need to be done without the benefit of vision. Doors and emergency exits may become jammed due to airframe distortion during an accident, or they may be too hard to open due to water pressure. Passengers should be aware that kicking out a window or the windshield may be the quickest and easiest way to exit the seaplane. Because many seaplanes come to rest in a nose-down position due to the weight of the engine, the baggage compartment door may offer the best path to safety.

In addition to covering these basic areas, be sure to tell passengers to leave everything behind in the event of a mishap except their PFD. Pilots should never assume that they will be able to assist passengers after an accident. They may be injured, unconscious, or impaired, leaving passengers with whatever they remember from the pilot's briefing. A thorough briefing with clear demonstrations can greatly enhance a passenger's chance of survival in the event of a mishap.

Chapter 9
Float and Ski Equipped Helicopters

Helicopters are capable of landing in places inaccessible to other aircraft. In addition to rooftops, mountain tops, pinnacles, and other unprepared locations, there are times when a pilot may have to operate a helicopter in areas that do not offer a solid place to land. For those operations, the normal skid gear configuration can be replaced with a set of floats for water operations or skis for winter operations.

Note: In this chapter, it is assumed that the helicopter has a counterclockwise main rotor blade rotation as viewed from above.

FLOAT EQUIPPED HELICOPTERS
Unlike airplanes, there is no additional rating required for helicopter float operations. However, it is strongly recommended that pilots seek instruction from a qualified instructor prior to operating a float equipped helicopter. Check the Pilot's Operating Handbook (POH) or Rotorcraft Flight Manual (RFM) for any limitations that may apply when operating with floats installed. [Figure 9-1]

Figure 9-1. Float equipped helicopter.

CONSTRUCTION AND MAINTENANCE
Helicopter floats are constructed of a rubberized fabric, or nylon coated with neoprene or urethane, and may be of the fixed utility or emergency pop-out type. Fixed utility floats typically consist of two floats that may have one or more individual compartments inflated with air. Fixed floats may be of the skid-on-float or the float-on-skid design. [Figure 9-2]

A **skid-on-float** landing gear has no rigid structure in or around the float. The float rests on the hard surface and supports the weight of the helicopter. With this

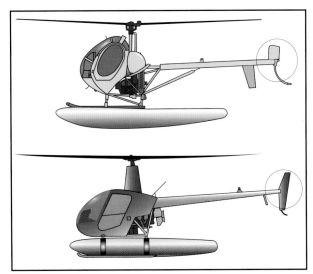
Figure 9-2. Skid-on-float and float-on-skid landing gear.

type of design, be aware of differences in float pressure. While the pressures are usually low, a substantial difference can cause the helicopter to lean while on a hard surface making it more susceptible to dynamic rollover.

A **float-on-skid** landing gear has modified skids that support the weight of the helicopter on hard surfaces. The floats are attached to the top of the skid and only support the weight of the helicopter in water. A float with low pressure or one that is completely deflated will not cause any stability problems on a hard surface.

Emergency **pop-out floats** consist of two or more floats with one or more individual compartments per float, depending on the size of the helicopter. [Figure 9-3] They are often inflated with compressed nitrogen

Figure 9-3. Pop-out float equipped helicopter.

or helium and are deployed prior to an emergency landing on water. The aircraft's maintenance manual states that the pop-out floats must be tested periodically through a deployment check, a leak check, and a hydro-static check of the compressed gas cylinder.

To maintain the floats in good condition, perform the following tasks before and after every flight:

• **Inflation**—Check each float compartment for proper inflation. Record the pressure to obtain a trend over time to help recognize leaks.

• **Condition**—Inspect the entire float assembly for cuts, tears, condition of chafing strips, and security of all components.

• **Clean**—Wash oil, grease, or gasoline from the floats, since they deteriorate the float's material.

• **Flush**—If the helicopter has been operated on salt water, flush the entire helicopter, including the float assembly, with plenty of fresh water.

• **Storage**—Avoid placing the floats in direct sunlight when not in use.

OPERATIONAL CONSIDERATIONS

Helicopter floats have only a mild effect on aircraft performance, with just a slight weight penalty and reduction in cruise speed. However, the large surface area of the floats makes the helicopter very sensitive to any departure from coordinated flight. For example, in cruise flight, any yawing causes the helicopter to roll in the opposite direction, as shown in figure 9-4. A failure of the engine requires immediate pedal application to prevent an uncontrollable yaw, with a resulting roll.

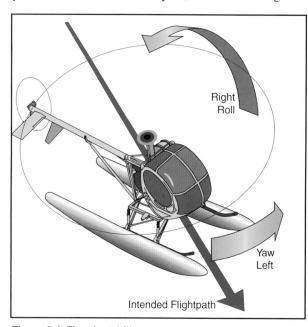

Right Roll

Yaw Left

Intended Flightpath

Figure 9-4. Float instability.

Similarly, a tail rotor failure in cruise flight requires immediate entry into autorotation to prevent a yaw and the subsequent roll. Corrections to this rolling moment can exceed rotor limits and cause mast bumping or droop stop pounding.

Helicopters equipped with skids-on-floats are limited in ground operations. Minimize horizontal movement during takeoffs and landings from hard surfaces to avoid scuffing or causing other damage to the floats. Perform approaches, in which hover power may not be available, by flaring through hovering altitude in a slightly nose-high attitude to reduce forward motion. Just prior to the aft portion of the floats touching down, add sufficient collective pitch to slow the descent and stop forward motion. Rotate the cyclic forward to level the helicopter, and allow the helicopter to settle to the ground, then reduce collective pitch to the full down position. In helicopters with low inertia rotor systems, an autorotation to a hard surface requires a more aggressive flare to a near-zero groundspeed to ensure minimal movement upon landing. A running takeoff or landing on a hard surface is not recommended in helicopters equipped with skids-on-floats.

Helicopters equipped with floats-on-skids are capable of performing running takeoffs and landings, and autorotations to hard surfaces require the same procedures as non-float equipped helicopters. The surfaces should be flat and clear of objects that may puncture, rip, or cause other damage to the floats. Do not attempt to land on the heels of floats-on-skids as they may cause the tail boom to kick up and be struck by the rotor.

Helicopters equipped with stored emergency pop-out floats are operated with the same procedures as a helicopter without floats. When emergency floats are deployed, the helicopter may have similar characteristics to a helicopter with fixed floats and should be flown accordingly. If emergency floats are deployed during autorotation, the increased surface increases parasite drag with a resulting reduction in airspeed. To regain the recommended autorotation airspeed, the nose must be lowered.

Effects on aircraft performance must also be considered during water operations. Air is often cooler near bodies of water, thus decreasing the density altitude but also increasing humidity. Although the higher humidity of the air has little effect on aerodynamic performance, it can reduce piston engine output by more then 10 percent. Properly leaning the mixture might possibly return some of this lost power.

Turbine engines experience only a small, often negligible, power loss in high humidity conditions.

Warning: During water operation, if there is any possibility that the tail rotor struck the water, do not attempt a takeoff. Although a tail rotor water strike may not show any visible evidence of damage, a tail rotor failure is likely to occur.

PREFLIGHT INSPECTION

The preflight inspection consists of the standard aircraft inspection with a few additional items associated with the floats. When performing a preflight inspection, follow the manufacturer's recommendations. A typical inspection of the floats includes:

- **Visual Inspection**—Examine the floats for cuts, abrasions, or other damage.

- **Inflation Check**—Although proper inflation can be checked by hand feeling for equal pressure and firmness, a pressure gauge is the preferred method to check for the correct pressure listed in the POH or RFM. For flights to higher altitudes, adjust float pressure before takeoff so that maximum pressure is not exceeded, unless the floats are equipped with pressure relief valves.

- **Valve Checks**—Check the air valves by filling the neck with water and watching for bubbles. Examine fittings for security and, if operated in salt water, inspect for corrosion.

- **Float Stabilizer,** if equipped—Examine the float stabilizer and other float related surfaces for security and condition. Any indication of water contact requires, at a minimum, a visual inspection of the tail surfaces, tail boom, and mounts. Consult the aircraft's maintenance manual for any additional required inspections.

- **Float and Skid Freedom**—In cold weather, it is common for floats and skids to freeze to the surface. Inspect the floats and skids for freedom of movement and obstructions. To help prevent this problem, try to park on a dry surface with proper drainage.

- **Secure**—Ensure all equipment is secure and properly stowed including survival equipment, anchors, tiedowns, and paddles. If possible, stow items inside the helicopter that could become loose and fly into the rotors.

- **Survival Equipment**—Check the quantity and condition of survival equipment including flotation devices, liferafts, provisions, and signaling devices.

STARTING

A helicopter on a hard surface has the friction of the skids or floats to counter the torque produced when the rotor is engaged. Therefore, you have more control over the helicopter if you can engage the rotors while it is sitting on a hard surface. On water, little or no anti-torque control is present until the rotor system has accelerated to approximately 50 percent of its normal operating r.p.m. A heavily loaded helicopter's floats sit deeper in the water and create more resistance to the turning force than a lightly loaded helicopter. Thus a helicopter turns less when heavily loaded and more when lightly loaded.

To overcome the spinning and to prevent drifting, tie the helicopter securely to a dock or to the shore using the fore and aft cross tubes if not otherwise indicated in the POH or RFM. If help is not available for casting off, it may be necessary to paddle to a clear location well away from the shoreline for a safe start. Wind and water currents may cause the helicopter to turn or drift a considerable distance before control is obtained. To compensate, use a starting position upwind and upcurrent of a clear area.

Illusions of movement or non-movement can make it difficult to maintain a fixed position during rotor engagement and runup. Techniques to overcome these illusions are discussed later.

TAXIING AND HOVERING

Where possible, it is usually more convenient and safer to hover taxi to the destination. However, due to power limits, local restrictions, noise, water spray, or creating a hazard to other vessels or people, it may be necessary to water taxi the helicopter. To taxi in water, maintain full rotor r.p.m. and use sufficient up collective to provide responsive cyclic control to move the helicopter. Never bottom the collective pitch while the helicopter is in motion to avoid momentarily sinking the floats or capsizing the helicopter. Float equipped helicopters should be taxied with the nose in the direction of movement. Maximum taxi speed is attained when the bow wave around the nose of the floats rises slightly above the normal waterline. Beyond this speed, the bow wave flows over the front portion of the floats, and this severe drag may capsize the helicopter. When the helicopter is heavily loaded, it is restricted to a slower taxiing speed than when lightly loaded. When taxiing in small waves, point the helicopter into or at a slight angle to the waves. Never allow the helicopter to roll in the trough. In some instances, increasing collective can produce enough downwash to create a slight smoothing effect on wind-produced waves.

A ground swell can be dangerous to the tail rotor while the helicopter is riding up and pitching over the swell.

Approach the swell at a 30° to 45° angle and use collective pitch to minimize bobbing. If it becomes obvious that continued water taxi could lead to a serious problem, lift the helicopter off and reassess the situation. It might be possible to land in an area that does not contain the same conditions.

When hovering over or taxiing on water, movement of the helicopter may be difficult to judge. The rippling effect of the water from the downwash makes it appear as if the helicopter is moving in one direction when it is in fact stationary or even moving in the opposite direction. To maintain a fixed position or maintain a straight course while taxiing and hovering, use a fixed reference such as the bank or a stationary object in the water. When reference points are not available, judge movement by swirls, burbles, or slicks seen around the floats.

Hovering a helicopter over open water can create deceptive sensations. Without a reference point, extensive or rapid helicopter movements may go unnoticed. Very smooth and very rough water aggravate this situation. The most desirable water conditions are moderate ripples from a light breeze. An odd sensation, similar to vertigo, is sometimes produced by the concentric outward ripples resulting from the rotorwash, and pilots must keep their eyes moving and avoid staring at any particular spot. The inexperienced pilot may choose to initiate a slight forward movement when taking off into or landing from a hover. This guards against undesirable backward or sideward drift during takeoff or landing. With smooth water conditions, the usual tendency is to hover too high because the outward-flowing ripples from the rotorwash gives the pilot the sensation of being in a bowl and descending.

TAKEOFF
A float equipped helicopter can perform a normal takeoff from a hover or directly from the water. If there is insufficient power available for a normal takeoff, a running takeoff from a slow forward taxi may be an option. However, remember that water creates drag, so with insufficient power, a running takeoff may not be possible either.

The preferred method for taking off from water is to move forward into translational lift without pausing to hover after leaving the water. This type of takeoff is similar to a normal takeoff from the surface.

A normal takeoff from a hover over water is similar to the same type of takeoff over a hard surface. A common problem is poor judgment of altitude and rate of acceleration, which causes the pilot to increase speed without an increase in altitude. This causes the helicopter to enter the high speed portion of the height/velocity diagram, reducing the probability of a successful autorotation in the event of an engine failure. Also, be aware of possible restricted visibility during takeoff from water spray produced by the rotors. To help alleviate these problem areas, as the helicopter begins to move forward, use reference points some distance in front of the helicopter.

Over water, ground effect is reduced from the absorption of energy in the downwash. This increases the power required to hover and with other factors may exceed the power available. When this occurs, perform a slow taxi to a takeoff to take advantage of the translational lift produced from the forward motion. Remember, translational lift is also affected by any wind that is present. Apply sufficient collective pitch to keep the floats riding high or skimming the surface. While skimming the surface, float drag increases rapidly, and the takeoff must be executed promptly since a further increase in speed, with the floats plowing in the water, is likely to exceed the limit of aft cyclic control or cause the floats to tuck under the water. The speed at which the floats tuck under is the maximum forward speed that can be attained and is determined by the load and attitude of the helicopter. Never lower the collective during this procedure because doing so could bury the nose of the floats in the water and possibly capsize the helicopter.

LANDING
Pilots performing glassy water landings may experience some difficulty in determining their altitude above the surface. The recommended procedure is to continue an approach to the surface with a slow rate of descent until making contact, avoiding any attempt to hover. The helicopter's downwash creates a disturbance in the water as concentric ripples moving away from the helicopter. Although this provides the pilot with a visual reference, it may also cause the sensation of moving backwards or descending rapidly. A natural tendency is to apply too much collective pitch in an attempt to halt the perceived descent. To overcome the effects of these visual illusions, avoid staring at the water near the helicopter and maintain forward and downward movement until contacting the water. When making approaches to a landing on a large body of water when land areas or other fixed objects are not visible, occasionally glance to either side of the horizon to avoid stare-fixation. Another technique some pilots use when fixed objects are not available, and the water is glassy, is to make a low pass over the area to create a disturbance on the surface. This disturbance remains for a while giving the pilot a reference to help determine distance.

When landing on water with a slight chop, bring the helicopter to a hover and descend vertically with no horizontal movement. This procedure is similar to landing on a hard surface.

Make a running landing on water when high density altitude or a heavy load results in insufficient power to hover. Perform this type of landing when sufficient power is not available to reduce the speed to 5 knots or less. When approaching with greater than 5 knots of speed, hold a slight nose-high attitude to allow the aft portion of the floats to plane. Maintain collective pitch until the speed reduces to below 5 knots, and the helicopter settles into the water. At zero groundspeed, slowly lower the collective into the full down position. Lowering the collective or leveling the helicopter too quickly may result in the floats tucking, which can cause the helicopter to capsize.

Caution: The following discussion deals with landing in heavy seas. Use these procedures only in an emergency.

Landing the float helicopter becomes **risky** when the height of short, choppy waves exceed one half the distance from the water to the helicopter's stinger, and the distance from crest to crest is nearly equal to or less than the length of the helicopter. These waves cause the helicopter to pitch rapidly and may bring the rotor blades in contact with the tail boom or the tail rotor in contact with the water. In addition, avoid landing parallel to steep swells as this could lead to dynamic rollover. [Figure 9-5]

If landing on waves higher than half the distance from water to stinger, the following techniques apply:

- Land the helicopter 30° to 45° from the direct heading into the swell. This minimizes the fore and aft pitching of the fuselage, reducing the possibility of the main rotor striking the tail boom, or the tail rotor contacting the water. This also minimizes the possibility of dynamic rollover.

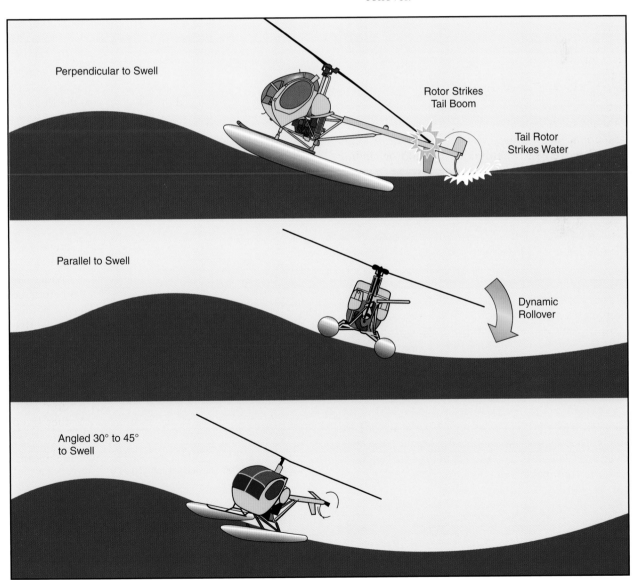

Figure 9-5. Effect of landing heading relative to waves.

- When landing with power, maintain rotor r.p.m. in the normal operating range. This permits a quick takeoff if the helicopter begins to pitch excessively or when an especially high wave becomes a hazard.

- When landing without power in high wave conditions, hold the desired heading as long as directional control permits. As the rotor r.p.m. decreases to the point that the desired heading cannot be maintained, bring the rotor to a stop as quickly as possible to avoid rotor contact with the tail boom.

AUTOROTATION

An autorotation to water is similar to one performed on a hard surface except that during touchdown, the helicopter is kept in a slight nose-high position. For greater safety, slow to around 5 knots of forward speed. However, if this is not possible, maintain a slight nose-high attitude and full-up collective to allow the floats to plane until the speed decelerates below 5 knots. As the helicopter settles to the surface and slows to zero knots, level the helicopter with cyclic and lower the collective. Do not lower the collective or level the helicopter until the speed has reduced sufficiently or the floats may tuck causing the helicopter to capsize. Hold a pitch attitude that keeps the tail from contacting the water.

Autorotations to smooth, glassy water may lead to depth perception problems. If possible, try to land near a shoreline or some object in the water. This helps in judging altitude just prior to touchdown.

SHUTDOWN AND MOORING

Although a helicopter can be moored prior to shutdown, it is preferable to fly to a landing spot on the dock or shore prior to shutting down. The helicopter can then be parked there. If mooring is the only option, be aware of any posts or pillars that might extend above the main dock level. Even though there may be plenty of blade clearance when the rotor is at full r.p.m., blade droop due to low r.p.m. could cause the blades to come into contact with items on the dock. Also be aware of wind and waves that could tilt the helicopter and cause the blades to contact objects. If near an ocean or large body of water, tides could change the water level considerably in just a few hours, so anticipate any changes and position the helicopter to prevent any damage due to the changing conditions.

When mooring the helicopter prior to shutting down, arrange the mooring lines so the tail cannot swing into objects once the rotors stop. Some pilots prefer to moor the helicopter nose in to protect the tail rotor.

If there is sufficient room to allow for drift and possible turning or weathervaning, the helicopter may be shut down on open water, but wind and water currents may move the helicopter a considerable distance. When shutting down on open water, do so upwind or upcurrent and allow the helicopter to drift to the mooring buoy or dock. It might be necessary to use a paddle to properly position the helicopter.

Because of the great danger from the main rotor or tail rotor of the helicopter to personnel, docks, or vessels, pilots should never attempt to water taxi up to a dock or vessel. In addition, loading or unloading passengers or freight from a partially afloat helicopter with the rotors turning is extremely dangerous. When loading or unloading passengers, the helicopter should be resting on a hard surface, either on the shore or on a helipad on a dock or on a boat. Passengers should always:

- stay away from the rear of the helicopter,

- approach or leave the helicopter in a crouching manner,

- approach from the side or front, but never out of the pilot's line of vision,

- hold firmly to loose articles and never chase after articles that are blown away by the rotor downwash, and

- never grope or feel their way toward or away from the helicopter.

GROUND HANDLING

On helicopters equipped with floats-on-skids, ground handling usually can be performed with normal or slightly modified ground handling wheels. With the ground handling wheels kept onboard, the helicopter can be handled at any landing facility. On helicopters equipped with skids-on-floats, the helicopter must be transported by a special dolly or wheeled platform on which the helicopter lands. Unless a dolly or platform is available at the destination, the aircraft usually remains where it lands.

SKI EQUIPPED HELICOPTERS

Ski equipped helicopters are capable of operating from snow and other soft surfaces that might otherwise inhibit conventional gear helicopters. [Figure 9-6] Snow can greatly reduce visibility causing pilot disorientation; therefore, special procedures are used when operating in snow.

Figure 9-6. Ski equipped helicopter.

CONSTRUCTION AND MAINTENANCE REQUIREMENTS

Helicopter skis are made from plastics and composite materials such as fiberglass with steel and aluminum hardware. Steel runners on the bottoms of the skis protect them during hard surface operations. Excessive wear of these runners can lead to wear or damage to the skis.

All of the steel bands securing the skis to the skids should have a protective rubber lining preventing the bands from wearing into the skids. This lining should be replaced if it becomes brittle or shows signs of wear.

Have any damage to the skis repaired before flight even if the skis are not needed, or simply have the skis removed. A cracked ski could break off and damage the helicopter or injure people on the ground.

OPERATIONAL CHARACTERISTICS

Apart from the small weight penalty and slight reduction in speed, a ski equipped helicopter operates exactly like one with no skis. The main concern when operating with skis is to avoid operations that may damage the skis, such as landing on rocks or rough hard surfaces.

PREFLIGHT REQUIREMENTS

The preflight inspection consists of the standard aircraft inspection and includes additional items associated with the skis. The POH or RFM contains the appropriate supplements and additional inspection criteria. Typical inspection criteria include:

- **Hardware**—Inspect all of the steel bands and bolts securing the skis to the skids for security. Check for any movement of the skis on the skids. A torque stripe can help determine if any movement has occurred.

- **Liner**—Inspect the rubber liner between the steel bands and the skids.

- **Runners**—Inspect the steel runners on the bottoms of the skis.

- **Condition**—Inspect the skis for cracks and check the edges for separation of fiber layers.

- **Clean**—Remove all snow and ice from the skis which could break off and cause damage to the tail rotor during flight.

- **Ski Freedom**—In cold weather, it is common for the skis to freeze to the surface. Inspect the skis for freedom of movement.

STARTING

Helicopter starting procedures on snow and ice are identical to a hard surface starting procedure except that care must be taken to maintain antitorque control on a slippery surface. When performing the free-wheeling unit check on ice, place the pedals in the autorotation position to prevent the helicopter from spinning.

TAXIING AND HOVERING

When hovering over snow, the rotorwash may create a white-out condition if sufficient loose snow is present. Blowing and drifting snow may give the illusion of movement in the opposite direction. When operating in snow, it is vital to select a reference point to maintain situational awareness and take off directly to a high hover at an altitude that allows visual contact to be maintained. When performing a hover taxi, select the speed just above effective translational lift to help keep the blowing snow behind the helicopter. If loose snow is less than 6 inches, it may be possible to apply collective pitch to create enough rotorwash to blow away the majority of the snow before lift-off. If moving the helicopter a short distance, and especially when around other aircraft, it might be preferable to surface taxi on the skis.

When taxiing wheel-equipped helicopters on snow and ice, use caution when applying the brakes. If the helicopter begins to skid sideways, lower the collective, which places all of the weight on the wheels and move the cyclic in the opposite direction of the skid. If the skid continues, the best option at that point is to bring the helicopter into a hover, but be aware of objects that could lead to a dynamic rollover situation.

TAKEOFF

Normal takeoff procedures are used in snow and ice, but before startup, check the departure path for any obstructions that may be obscured by blowing snow. Powerlines are difficult to see in the best conditions and nearly impossible to recognize through blowing snow.

Perform a takeoff from a hover or from the surface by fairly quickly increasing speed through effective translational lift and gaining altitude in order to fly out of the low visibility conditions. A takeoff from ice requires slow application of power and proper pedal application to prevent spinning. At certain temperatures, the skis may freeze to ice surfaces. If this occurs, a slight left and right yawing with the pedals may break the helicopter free. If this does not free the skids, shut down the helicopter and free them manually. Excessive pedal application could damage the skids.

LANDING

As with takeoffs, landings in snow can prove to be extremely hazardous if reference points are not available. When possible, land near objects that won't be easily obscured by blowing snow. If none are available, drop a marker made from a heavy object, such as a rock tied to a colored cloth; then retrieve it after landing.

When the snow condition is loose or unknown, make a zero-groundspeed landing directly to the surface without pausing to hover. A shallow approach and running landing can be performed when the snow is known to be hard packed and obstacles are not hidden under the snow. The lower power required in a running landing reduces the downwash and the forward motion keeps blowing snow behind the helicopter until after surface contact.

If the surface conditions are unknown, a low reconnaissance flight might be appropriate. This could be followed by a low pass. A low pass might blow away loose snow and keep the debris behind the helicopter. If the surface appears appropriate for a landing, make an approach to a high hover to blow away any remaining

loose snow and begin a vertical descent to the landing. If the surface appears to be deep hard-packed snow or ice, lower the collective slowly on landing and watch for cracking in the surface. Should one skid break through the surface, a dynamic rollover is likely to follow, so be prepared to return to a hover if the surface is unstable.

Skis are also very useful for landing on uneven or soft, spongy surfaces. They provide a larger surface area to support the helicopter, thus assisting in stability. Be sure that the skis are not hooked under roots or brush during lift-off.

AUTOROTATION

Use normal autorotation procedures in ski equipped helicopters. Perform practice autorotations on snow or sod to reduce the wear on the skis.

GROUND HANDLING

Shut down before loading and unloading. If shutting down is not feasible, load and unload passengers only from the front during snow and ice operations. This prevents the main rotors from striking an individual should one landing gear drop through the snow or ice. Beware of loading and unloading while running in deep snow as the rotor clearance is reduced by the height of the snow above the skids.

Most skis for skid-equipped helicopters allow use of standard or slightly modified ground handling wheels. Skis for wheel-equipped helicopters often have cutouts to allow the wheels to protrude slightly below the ski for ground handling.

Glossary

AMPHIBIAN—A seaplane with retractable wheel-type landing gear that can be extended to allow landings to be made on land.

ANCHOR—A heavy hook connected to the seaplane by a line or cable, intended to dig into the bottom and keep the seaplane from drifting.

AUXILIARY FIN — An additional vertical stabilizer installed on some float planes to offset the increased surface area of the floats in front of the center of gravity.

BEACHING—Pulling a seaplane up onto a suitable shore so that its weight is supported by relatively dry ground rather than water.

BEAUFORT WIND SCALE—A standardized scale ranging from 0-12 correlating the velocity of the wind with predictable surface features of the water.

BILGE—The lowest point inside a float, hull, or watertight compartment.

BILGE PUMP—A pump used to extract water that has leaked into the bilge of a float or flying boat.

BULKHEAD—A structural partition that divides a float or a flying boat hull into separate compartments and provides additional strength.

BUOYANCY—The tendency of a body to float or to rise when submerged in a fluid.

BUOYS—Floating objects moored to the bottom to mark a channel, waterway, or obstruction.

CAN BUOYS— Cylindrical buoys marking the left side of a channel for an inbound vessel. They have odd numbers which increase from seaward.

CAPSIZE—To overturn.

CAST OFF—To release or untie a vessel from its mooring point.

CENTER OF BUOYANCY—The average point of buoyancy in floating objects. Weight added above this point will cause the floating object to sit deeper in the water in a level attitude.

CHINE—The longitudinal seam joining the sides to the bottom of the float. The chines serve a structural purpose, transmitting loads from the bottoms to the sides of the floats. They also serve a hydrodynamic purpose, guiding water away from the float, reducing spray, and contributing to hydrodynamic lift.

CHOP— A roughened condition of the sea surface caused by local winds. It is characterized by its irregularity, short distance between crests, and whitecaps.

COMBINATION SKI— A type of aircraft ski that can be used on snow or ice, but that also allows the use of the skiplane's wheels for landing on runways.

CREST—The top of a wave.

CURRENT — The horizontal movement of a body of water.

DAYBEACONS — Unlighted beacons.

DAYMARKS—Conspicuous markings or shapes that aid in making navigational aids readily visible and easy to identify against daylight viewing backgrounds.

DECK—The top of the float, which can serve as a step or walkway. Bilge pump openings, hand hole covers, and mooring cleats are typically located along the deck.

DISPLACEMENT POSITION—The attitude of the seaplane when its entire weight is supported by the buoyancy of the floats, as it is when at rest or during a slow taxi. Also called the idling position.

DOCK—To secure a seaplane to a permanent structure fixed to the shore. As a noun, the platform or structure to which the seaplane is secured.

DOWNSWELL—Motion in the same direction the swell is moving.

FETCH—An area where wind is generating waves on the water surface. Also the distance the waves have been driven by the wind blowing in a constant direction without obstruction.

FLOATPLANE — A seaplane equipped with separate floats to support the fuselage well above the water surface.

FLOATS—The components of a floatplane's landing gear that provide the buoyancy to keep the airplane afloat.

FLOATS-ON-SKIDS—A type of helicopter float design where the floats sit on top of the fully functional skids. During water operations, the floats support the weight of the aircraft, and on hard surfaces the skids support the weight of the aircraft.

FLYING BOAT—A type of seaplane in which the crew, passengers, and cargo are carried inside a fuselage that is designed to support the seaplane on the water. Also called a hull seaplane.

GLASSY WATER—A calm water surface with no distinguishable surface features, with a glassy or mirror like appearance. Glassy water can deceive a pilot's depth perception.

HEIGHT-TO-LENGTH RATIO—The ratio between the height of a swell to the length between two successive crests (swell length).

HYDRODYNAMIC FORCES—Forces relating to the motion of fluids and the effects of fluids acting on solid bodies in motion relative to them.

HYDRODYNAMIC LIFT—For seaplanes, the upward force generated by the motion of the hull or floats through the water. When the seaplane is at rest on the surface, there is no hydrodynamic lift, but as the seaplane moves faster, hydrodynamic lift begins to support more and more of the seaplane's weight.

IDLING POSITION—The attitude of the seaplane when its entire weight is supported by the buoyancy of the floats, as it is when at rest or during a slow taxi. Also called the displacement position.

KEEL—A strong longitudinal member at the bottom of a float or hull that helps guide the seaplane through the water, and, in the case of floats, supports the weight of the seaplane on land.

LEEWARD—Downwind, or the downwind side of an object.

MOOR—To secure or tie the seaplane to a dock, buoy, or other stationary object on the surface.

NUN BUOYS—Conical buoys marking the left side of a channel for an inbound vessel. They often have even numbers that increase as the vessel progresses from seaward.

PLAIN SKI—A type of aircraft ski that can only be used on snow or ice, as compared to combination skis, which also allow the use of the skiplane's wheels for landing on runways.

PLANING POSITION—The attitude of the seaplane when the entire weight of the aircraft is supported by hydrodynamic and aerodynamic lift, as it is during high-speed taxi or just prior to takeoff. This position produces the least amount of water drag. Also called the step position, or "on the step."

PLOWING POSITION—A nose high, powered taxi characterized by high water drag and an aftward shift of the center of buoyancy. The weight of the seaplane is supported primarily by buoyancy, and partially by hydrodynamic lift.

POP-OUT FLOATS—Helicopter floats that are stored deflated on the skids or in compartments along the lower portion of the helicopter, and deployed in the event of an emergency landing on water. Compressed nitrogen or helium inflates the floats very quickly.

PORPOISING—A rhythmic pitching motion caused by an incorrect planing attitude during takeoff.

PORT—The left side or the direction to the left of a vessel.

PRIMARY SWELL—The swell system having the greatest height from trough to crest.

RAMPING—Using a ramp that extends under the water surface as a means of getting the seaplane out of the water and onto the shore. The seaplane is typically driven under power onto the ramp, and slides partway up the ramp due to inertia and engine thrust.

SAILING—Using the wind as the main motive force while on the water.

SEA—Waves generated by the existing winds in the area. These wind waves are typically a chaotic mix of heights, periods, and wavelengths. Sometimes the term refers to the condition of the surface resulting from both wind waves and swells.

SEA STATE CONDITION NUMBER—A standard scale ranging from 0-9 that indicates the height of waves.

SEAPLANE — An airplane designed to operate from water. Seaplanes are further divided into flying boats and floatplanes.

SEAPLANE LANDING AREA—Any water area designated for the landing of seaplanes.

SEAWARD—The direction away from shore.

SECONDARY SWELLS—Those swell systems of less height than the primary swell.

SISTER KEELSONS—Structural members in the front portion of floats lying parallel to the keel and midway between the keel and chines, adding structural rigidity and adding to directional stability when on the water.

SKEG—A robust extension of the keel behind the step which helps prevent the seaplane from tipping back onto the rear portion of the float.

SKIDS-ON-FLOATS—A type of helicopter float design where the rigid portion of the landing gear rests on the floats. The floats support the whole weight of the helicopter in water or on hard surfaces.

SKIPPING — Successive sharp bounces along the water surface caused by excessive speed or an improper planing attitude when the seaplane is on the step.

SPONSONS—Short, winglike projections from the sides of the hull near the waterline of a flying boat. Their purpose is to stabilize the hull from rolling motion when the flying boat is on the water, and they may also provide some aerodynamic lift in flight. Tip floats also are sometimes known as sponsons.

SPRAY RAILS—Metal flanges attached to the inboard forward portions of the chines to reduce the amount of water spray thrown into the propeller.

STARBOARD—The right side or the direction to the right of a vessel.

STEP—An abrupt break in the longitudinal lines of the float or hull, which reduces water drag and allows the pilot to vary the pitch attitude when running along the water's surface.

STEP POSITION—The attitude of the seaplane when the entire weight of the aircraft is supported by hydrodynamic and aerodynamic lift, as it is during high-speed taxi or just prior to takeoff. This position produces the least amount of water drag. Also called the planing position.

SWELL—Waves that continue after the generating wind has ceased or changed direction. Swells also are generated by ships and boats in the form of wakes, and sometimes by underwater disturbances such as volcanoes or earthquakes. The waves have a uniform and orderly appearance characterized by smooth, rounded, regularly spaced wave crests.

SWELL DIRECTION — The direction from which a swell is moving. Once set in motion, swells tend to maintain their original direction for as long as they continue in deep water, regardless of wind direction. Swells may be moving into or across the local wind.

SWELL FACE—The side of the swell toward the observer. The back is the side away from the observer. These terms apply regardless of the direction of swell movement.

SWELL LENGTH—The horizontal distance between successive crests.

SWELL PERIOD — The time interval between the passage of two successive crests at the same spot in the water, measured in seconds.

SWELL VELOCITY — The velocity with which the swell advances with relation to a fixed reference point, measured in knots. There is little movement of water in the horizontal direction. Each water particle transmits energy to its neighbor, resulting primarily in a vertical motion, similar to the motion observed when shaking out a carpet.

TIDES—The alternate rising and falling of the surface of the ocean and other bodies of water connected with the ocean. They are caused by the gravitational attraction of the sun and moon occurring unequally on different parts of the earth. Tides typically rise and fall twice a day.

TIP FLOATS—Small floats near the wingtips of flying boats or floatplanes with a single main float. The tip floats help stabilize the airplane on the water and prevent the wingtips from contacting the water.

TRANSOM—As it applies to seaplanes, the rear bulkhead of a float.

TROUGH—The low area between two wave crests.

UPSWELL—Motion opposite the direction the swell is moving. If the swell is moving from north to south, a seaplane going from south to north is moving upswell.

VESSEL—Anything capable of being used for transportation on water, including seaplanes.

WATER RUDDERS—Retractable control surfaces on the back of each float that can be extended downward into the water to provide more directional control when taxiing on the surface. They are attached by cables and springs to the air rudder and operated by the rudder pedals in the cockpit.

WEATHERVANING—The tendency of an aircraft to turn until it points into the wind.

WINDWARD—Upwind, or the upwind side of an object.

WING FLOATS—Stabilizer floats found near the wingtips of flying boats and single main float floatplanes to prevent the wingtips from contacting the water. Also called tip floats.

Index

A

Aids for marine navigation 1-2
Altimeter setting 6-7
Amphibians 2-1, 6-2
Anchoring 6-9
Autorotation 9-2, 9-6, 9-8
Auxiliary fin 2-4, 5-2

B

Beaching 6-8, 6-10
Bilge pump 4-2
Bilge pump openings 2-2, 4-2
Bulkheads, float 2-2
Buoyancy 2-2, 4-3
Buoys 1-2, 1-3, 1-4

C

Center of buoyancy 4-4, 4-6,
Center of gravity 4-1, 5-1, 5-2, 5-3, 7-7
Centrifugal force (in turns) 4-6, 4-7, 4-14
Certificate, limitations 1-1
Chine 2-2
Clamp-on ski 7-1
Coast Guard rules 1-2
Combination ski 7-1, 7-2
Confined area operations 4-16, 6-7
Corrosion 4-1, 4-3
Crosswind 4-12, 4-13, 6-3, 7-5
Current 3-2, 4-8, 4-9, 6-5

D

Daybeacons and daymarks 1-2, 1-3, 1-4
Deck 2-2
Density altitude 4-11, 4-12, 5-1, 6-8, 9-5
Displacement 2-2, 4-3
Displacement position 4-3

Displacement
of float 2-2
position or attitude 4-3, 4-10
taxi 4-3
Docking 6-8, 6-10
Downwind takeoff 4-14

E

Escaping a submerged seaplane 8-8

F

Fetch 3-2, 8-1
Float construction 2-2, 2-3, 9-1
Float, weight-bearing capability 2-2, 9-1
Floatplane defined 2-1
Flying boat
definition 2-1
handling 4-9, 5-3

G

Glaciers 7-6
Glassy water 3-3, 4-15, 6-5, 9-4
Go-around 6-2, 6-8

H

Hovering 9-3, 9-7
Hull 2-1, 5-3
Hump (water drag) 4-9, 4-10, 4-11
Hydrodynamic lift 2-2, 4-4, 4-10

I

Ice (in floats) 4-3
Ice types 7-2
Idling 4-3, 4-8
Inland waters 1-2
International waters 1-2

K

Keel 2-2

L

Landings
    confined area 6-7
    crosswind 6-3
    downwind 6-5, 8-4
    emergency 6-8, 7-8
    frozen lakes and rivers 7-6
    glaciers 7-6
    glassy water 6-5
    helicopter 9-4, 9-8
    night landing 6-8, 7-8, 8-5, 8-6
    normal 6-3
    open sea 8-1
    rough water 6-7, 8-1, 9-5
    skiplane 7-6, 7-7
    tundra 7-6
Launching 4-3
Lighting conditions 7-6
Limitations of sea rating 1-1

M

Marine aids for navigation 1-2
Mooring 6-8, 6-9, 9-6

N

Night operations 6-8, 8-5, 8-6
Noise 3-4, 4-12, 6-2
Normal takeoff 4-12

O

On the step 4-4, 6-2

P

Parking 7-7
Passenger briefing 4-3
Penetration ski 7-2
Plain ski 7-1
Planing position 4-4

Plow turn 4-6, 4-7
Plowing position 4-4
Pop-out floats 9-1
Porpoising 4-9, 5-3
Preflight inspection
    float equipped helicopter 9-3
    seaplane 4-1
    skiplane 7-3
    ski equipped helicopter 9-7
Privileges and Limitations 1-1

R

Ramping 6-8, 6-10
Regulations 1-1
Retractable ski 7-1
Right-of-way rules 1-2
Roll-on ski 7-1
Rough water 4-16, 6-7, 8-1, 9-5
Rules of the Sea 1-2
Runup 4-12
Runup (skiplane) 7-4

S

Sailing 4-8, 4-9
Seaplane defined 2-1
Seaplane landing areas
    beacons 1-2
    chart symbols 1-2
    reconnaissance 6-1
    restrictions 3-4
    unplanned 5-2
Sister keelsons 2-2
Skeg 2-2, 2-4
Ski types 7-1
Skids-on-floats 9-1, 9-6
Skipping 4-10
Snow types 7-2
Sponson 2-1
Spray damage 4-1
Spray rail 2-2, 4-2
Starting
    helicopter 9-3, 9-7
    seaplane 4-3
    skiplane 7-4
Step 2-3, 4-4
Step position 4-4
Step taxi 4-5, 6-3
Step turns 4-7
Survival equipment 7-3, 7-4, 7-8
Swell 3-2, 4-9, 6-2, 6-7, 8-1, 8-2, 8-3, 8-4, 8-5